Environment and Tourism

For many people 'going on holiday' is an increasingly central feature of contemporary western society. The tourism industry has expanded rapidly since 1950, but are environments benefiting from or being damaged by the tourists who visit them?

Environment and Tourism is an introductory text about the relationship that exists between tourism, society and the environment. The book examines the meanings of 'tourism' and 'environment' and gives a historical overview of the growth of tourism. It discusses how the tourism industry markets physical and cultural environments, to sell into the consumer market. Inevitably there have been consequences from the growth of tourism on environments and the use of environmental management and planning techniques is described. Ethical dimensions of the interaction between tourism and the environment are also considered. The book suggests ways in which the economics of tourism can be adopted in a positive way to aid conservation, in light of the failure of conventional economics to deal with problems of resource usage and pollution.

Environment and Tourism proceeds to look at whether the concept of sustainability can be applied to tourism and provides a critique of the 'new' forms of tourism that developed towards the end of the twentieth century. An extensive range of international case studies is used to illustrate the theoretical ideas presented. To aid the student there are chapter outlines, end-of-chapter summaries and further reading sections.

Andrew Holden is Principal Research Fellow in Tourism at the University of Luton.

Routledge Introductions to Environment Series
Published and Forthcoming Titles

Routledge Introductions to Environment Series

Environment and Tourism

Andrew Holden

London and New York

First published 2000
by Routledge
11 New Fetter Lane, London EC4P 4EE

Simultaneously published in the USA and Canada
by Routledge
29 West 35th Street, New York, NY 10001

Reprinted 2001

Routledge is an imprint of the Taylor & Francis Group

Typeset in Times by Keystroke, Jacaranda Lodge, Wolverhampton
Printed and bound in Great Britain by St Edmundsbury Press,
Bury St Edmunds, Suffolk

British Library Cataloguing in Publication Data
A catalogue record for this book is available from the British Library

Library of Congress Cataloging in Publication Data
Holden, Andrew.
      Environment and tourism / Andrew Holden.
         p. cm. – (Routledge introductions to environment series)
      Includes bibliographical references (p.).
       1. Tourism–Environmental aspects. 2. Tourism–Economic aspects.
   3. Tourism–Social aspects. I. Title: Data sheet: Environment & tourism.
  II. Title. III. Series.

G155.A1 H64 2000
338.4′791–dc21   00–038250

ISBN 0–415–20717–7 (hbk)
ISBN 0–415–20718–5 (pbk)

Contents

Series editor's preface
Environment and Society titles

The modern environmentalist movement grew hugely in the last third of the twentieth century. It reflected popular and academic concerns about the local and global degradation of the physical environment which was increasingly being documented by scientists (and which is the subject of the companion series to this, Environmental Science). However it soon became clear that reversing such degradation was not merely a technical and managerial matter: merely knowing about environmental problems did not of itself guarantee that governments, businesses or individuals would do anything about them. It is now acknowledged that a critical understanding of socio-economic, political and cultural processes and structures is central in understanding environmental problems and establishing environmentally sustainable development. Hence the maturing of environmentalism has been marked by prolific scholarship in the social sciences and humanities, exploring the complexity of society-environment relationships.

Such scholarship has been reflected in a proliferation of associated courses at undergraduate level. Many are taught within the 'modular' or equivalent organisational frameworks which have been widely adopted in higher education. These frameworks offer the advantages of flexible undergraduate programmes, but they also mean that knowledge may become segmented, and student learning pathways may arrange knowledge segments in a variety of sequences – often reflecting the individual requirements and backgrounds of each student rather than more traditional discipline-bound ways of arranging learning.

The volumes in this Environment and Society series of textbooks mirror this higher educational context, increasingly encountered in the early twenty-first century. They provide short, topic-centred texts on social science and humanities subjects relevant to contemporary society-environment relations. Their content and approach reflect the fact that each will be read by students from various disciplinary backgrounds, taking in not only social sciences and humanities but others such as physical and natural sciences. Such a readership is not always familiar with the disciplinary background to a topic, neither are readers necessarily going on to further develop their interest in the topic. Additionally, they cannot all automatically be thought of as having reached a similar stage in their studies – they may be first-, second- or third-year students.

The authors and editors of this series are mainly established teachers in higher education. Finding that more traditional integrated environmental studies and specialised texts do not always meet their own students' requirements, they have often had to write course materials more appropriate to the needs of the flexible undergraduate programme. Many of the volumes in this series represent in modified form the fruits of such labours, which all students can now share.

Much of the integrity and distinctiveness of the Environment and Society titles derives from their characteristic approach. To achieve the right mix of flexibility, breadth and depth, each volume is designed to create maximum accessibility to readers from a variety of backgrounds and attainment. Each leads into its topic by giving some necessary basic grounding, and leaves it usually by pointing towards areas for further potential development and study. There is introduction to the real-world context of the text's main topic, and to the basic concepts and questions in social sciences/humanities which are most relevant. At the core of the text is some exploration of the main issues. Although limitations are imposed here by the need to retain a book length and format affordable to students, some care is taken to indicate how the themes and issues presented may become more complicated, and to refer to the cognate issues and concepts that would need to be explored to gain deeper understanding. Annotated reading lists, case studies, overview diagrams, summary charts and self-check questions and exercises are among the pedagogic devices which we try to encourage our authors to use, to maximise the 'student friendliness' of these books.

Hence we hope that these concise volumes provide sufficient depth to maintain the interest of students with relevant backgrounds. At the same time, we try to ensure that they sketch out basic concepts and map their territory in a stimulating and approachable way for students to whom the whole area is new. Hopefully, the list of Environment and Society titles will provide modular and other students with an unparalleled range of perspectives on society-environment problems: one which should also be useful to students at both postgraduate and pre-higher education levels.

David Pepper
May 2000

Series International Advisory Board

Australasia: Dr P. Curson and Dr P. Mitchell, Macquarie University

North America: Professor L. Lewis, Clark University; Professor L. Rubinoff, Trent University

Europe: Professor P. Glasbergen, University of Utrecht; Professor van Dam-Mieras, Open University, The Netherlands

Figures

Boxes

Preface

This book forms part of a series produced by Routledge on the theme of 'environment and society'. The particular emphasis of this publication is upon the interaction that exists between the environment and tourism. Its main purpose is to provide an introductory text for undergraduate students to the concepts and themes that govern this interaction.

Two simple words, 'tourism' and 'environment', but in the following 70,000 words it is only possible to introduce the reader to the complex issues that lie behind these two terms. Both 'tourism' and the 'environment' represent complicated concepts, and both can be interpreted as intricate systems, where actions taken in one part of the system have consequences for its other component parts. This book takes an holistic approach, in trying to understand the complexity of both tourism and the environment, and the relationship that exists between them. Consequently, the book adopts a multidisciplinary stance to the investigation of this relationship and should be of interest to a range of students studying under the aegis of social science. The perspectives used in this book come from the disciplines of geography, sociology, social psychology and economics, as well as the fields of environment, development and tourism studies.

The term 'environment' is used in its widest sense to incorporate all aspects of human behaviour. Cultural, political, economic and social aspects of the environment are an important part of this book, besides purely considerations of the physical environment. All of these factors affect the way we live, and how as humans we interact with each other, as well as the non-human world. In the last half of the twentieth century the world has witnessed a faster pace of economic development than ever before. This pace of development has placed a tremendous strain upon the natural resources of the earth, and the functioning of the environmental systems that we as humans rely upon for our survival.

Global warming, ozone depletion, desertification and acid rain are all examples of the types of negative environmental changes that have resulted from, or been accelerated by, human actions. Humans' rights are also sometimes threatened by development, as humans continue to deny rights to other humans, and people are displaced from their traditional lands and denied access to resources. Subsequently, our interaction with fellow humans and the non-human world presents us with many ethical questions. At the beginning of the twenty-first century, we are increasingly being forced to face ethical issues, if not for the survival of other humans and species then for our own survival. The realisation that we are part of and not separate from this system, which we call 'environment', is beginning to show signs of dawning on an increasing number of people.

Changes in society, which can be traced to the Industrial Revolution almost two hundred years ago, are also influencing the way we live. Increasing rates of urbanisation in both the developed and less developed worlds, and the adoption of an ideology of consumerism as a global creed, are placing increasing demands upon our environment to satisfy our needs and desires. The process of urbanisation has had the effect of removing people from nature, and has presented people with the need to define new notions of community. These changes have developed needs and other wants which, combined with an increased level of prosperity, are increasingly satisfied through consumerism. One form of consumerism, which seems to be an increasingly popular way to meet these needs, is tourism. At the beginning of this century, there are over 650 million people travelling internationally on an annual basis, which is expected to rise to 1,600 million in twenty year's time. For many in the economically advanced societies of the West, tourism has therefore become a necessity of life, an experience that is consumed with increasing enthusiasm.

A trend of this expanding demand for tourism is for tourists to go further and further afield from where they live. Packages have already been sold to tourists for the first flights to space which are expected to take place within the next five years. Many of the world's coastlines and mountain areas have been developed for tourism and even Antarctica is now a part of the tourist menu. In Spain, the first country to experience mass international tourism in the 1950s, tourism has brought vast economic, cultural and physical environmental changes. These changes are both positive and negative; as will be seen in this book, tourism can be an agent of both positive and negative changes. Recognition of tourism as an agent of change – its growing importance in the world economy and the effects that it can have for the environment – was given at the United Nations General Assembly Special Session (Earth Summit II) in New York in 1997. The basis of this recognition was that tourism must be developed in a sustainable fashion, to ensure the conservation of resources for future generations to make their livelihoods from tourism, just as their parents do now. A fine aim that probably few of us would disagree with, but in reality a concept that represents a diversity of opinion over how it should be achieved, which in turn reveals much about the complexity of the environment in which tourism operates. It is this complexity of the interaction between 'tourism' and the 'environment' that this book sets out to explore.

Acknowledgements

I would like to thank the following people directly for either the help or advice they have given me during writing this book. For the information and interviews: Hilary Robinson, Thomson Travel Group; Lotta Sand, Fritidsresor; Piek Martin, Fritidsresor; Judy Smith-Spala, Grecotels; Andy Martin, Market Opinion Research Institute (MORI); and Dr Wolf Iwand of Touristik Union International. I would also like to give special thanks to Mike Stabler of the University of Reading and John Curran of the University of North London for their advice with Chapter 4. Thanks also to Tourism Concern, for the permission to reprint the Himalayan code shown in Box 5.14, to Butterworth-Heinemann Publishers for the permission to reprint Figure 7.2, and Her Majesty's Stationery Office for the permission to reprint Figures 2.1 and 2.2. Thanks also to Ann Michael of Routledge for advice on the practicalities of this book, and to the three reviewers of the draft manuscript, whose positive criticisms were helpful in putting together the final copy.

My general thanks go to all the colleagues and students I have had the pleasure of working with over the last decade. The list of people I would like to credit would run into pages. However, I would like to give special thanks to the following, for making working in the field of tourism studies both intellectually stimulating and fun, though not necessarily always simultaneously: Dr Raoul Bianchi, Robert Cleverdon, Patrick Coghlan, Professor Graeme Evans, Professor David Harrison, Professor Michael Hitchcock, Claire Kitts, Nikki MacCleod, Dr Julie Scott, Steve Shaw, John Sparrowhawk, Dr Marcus Stephenson, Dr Judy White and Professor Tom Selwyn, all at the University of North London; Professor Brian Goodall at the University of Reading; Kaija Lindroth at Helia Eastern Uusimaa Polytechnic; Dr Bill Bramwell, John Swarbrooke and Diane Laws at Sheffield Hallam University; and Professor Peter Burns, Dr Keith Hollinshead and Dr Peter Mason at the University of Luton. Finally, but by no means last, my thanks goes to my partner Kiran Kalsi for not complaining too much about all the late nights!

1 Introducing tourism

- Understanding tourism
- What tourism is and what it is not
- The history of tourism
- Conditions for the growth of tourism

Introduction

To understand the interaction that exists between tourism and the environment it is necessary to understand the complexity of tourism. Tourism is not just something that occurs in the environments of destinations overseas but is a function of the interaction of different factors in contemporary society. Since the 1950s there has been a rapid increase in the demand in western societies for people to travel internationally and visit a variety of different destinations. This growing demand for tourism is a reflection of changing economic and social conditions in our home environment, as much as it is about the physical and cultural characteristics of the environments that await tourists in other countries. This chapter examines the meaning and complexity of tourism, its history, and how social changes since the Industrial Revolution have shaped contemporary mass participation in tourism.

What is meant by tourism?

The acceptance of 'going away' on holiday, commonly referred to as tourism, as a part of our lifestyle in contemporary western society may lead us to believe that it has always been a feature of people's lives. Yet the word 'tourist' is a fairly new addition to the English language, the word 'tour-ist' (deliberately hyphenated), first appearing in the early nineteenth century (Boorstin, 1961). Boorstin draws a distinction between the arduous conditions undertaken by 'travellers' (a term originating from the French word *travail* meaning work, trouble, torment), such as pilgrims, and the 'tourist', for whom travel has become an organised and packaged affair.

The idea of travel for pleasure, for example to visit beautiful landscapes as opposed to travel for necessity or to demonstrate religious piousness, is therefore within the context of human activity a relatively recent phenomenon. Until the nineteenth century travel was not an easy option, nor were landscapes that we now regard as aesthetically pleasing, necessarily regarded in the same way. Yet, today the word 'tourism' has become part of our common language, with over 650 million people travelling internationally at the beginning of the twenty-first century (World Tourism Organisation, 1998a).

Despite this impressive figure, trying to define what tourism actually is has proved to be more problematic than might be expected. This difficulty is a reflection of both the complexity of tourism, and the fact that different stakeholders or groups with an interest in tourism are likely to have different aspirations of what they hope to achieve from it, and subsequently hold different perspectives on what it means to them.The stakeholders in tourism include governments, the tourism industry, local communities and tourists.

Definitions and types of tourism

For the majority of people who possess the financial means to participate in travel for recreational purposes, tourism is an activity that probably little conscious thought is given to beyond recollecting the enjoyment of the last holiday, and deciding where to go to for the next one. Yet this seemingly simple process involves the participation of national governments, tourism businesses and local communities, all of whom will have their own interests in tourism, which often leads them to be referred to as stakeholders in tourism.

Attempts to define tourism are made difficult because it is a highly complicated amalgam of various parts. These parts are a diverse range of factors, including the following: human feelings, emotions and desires; attractions built upon natural and developed resources; suppliers of transport, accommodation, and other services; and government policy and regulatory frameworks. Subsequently it is difficult to arrive at a consensual definition of what tourism actually is. Many authors of tourism texts (for example Mathieson and Wall, 1982; Murphy, 1985; Middleton 1988; Bull, 1991; Laws, 1991; Ryan 1991; Mill and Morrison, 1992; Davidson, 1993; Gunn, 1994; Burns and Holden, 1995; Cooper *et al.*, 1998; and Holloway, 1998) comment upon the problem of defining tourism.

Yet trying to understand the meaning of 'tourism' is important if we are to plan the use of natural resources and manage impacts associated with its development. What all commentators would probably agree with is that tourism involves travel, although how far one has to travel and how long one has to be away from one's home location to be categorised as a tourist, is debatable. A convenient definition that overcomes this difficulty is the one proposed by the World Tourism Organisation (1991) which was subsequently endorsed by the UN Statistical Commission in 1993: 'Tourism comprises the activities of persons travelling to and staying in places outside their usual environment for not more than one consecutive year for leisure, business or other purposes.'

The preceding definition challenges the commonly held perception that tourism is purely concerned with recreation and having fun. Whilst recreational tourism is the most usual form of tourism other types of tourism also exist. For instance Davidson (1993) besides recognising leisure or recreation (in which he includes travel for holidays, sports, cultural events, and visiting friends and relatives) as the main type of tourism, draws attention to the point that people also travel for business, study (or education), religious and health purposes. Indeed the origins of tourism lie in travel for reasons of faith, education and health, as is detailed later in this chapter.

Although business tourism may initially seem to have little relevance to a text dealing with the interaction between tourism and the environment, it is a particularly important market sector for the economies of many urban environments. Tourism has purposefully been used in government policy as a catalyst to aid the regeneration of economically depressed areas of post-industrial cities, such as Baltimore in the USA and Liverpool in the United Kingdom, a theme discussed in Chapter 3. Business travel can be viewed as including travel for the purposes of commerce, exhibitions and trade fairs, and conferences.

From the previously cited World Tourism Organisation (1991) definition it can inferred that tourism involves some element of interaction with a different type of environment to the one found at home. The consequences of this interaction are commonly referred to as the 'impacts of tourism', and can be categorised into the three main types, economic, social and environmental. All of these types of impacts can be either positive or negative and are discussed more fully in the course of this book. Recognition of the impacts that tourism can have on a destination environment are noted in the following definition of tourism given by Mathieson and Wall (1982: 1): 'The study of tourism is the study of people away from their usual habitat, of the establishments which respond to the requirements of travellers, and of the impacts that they have on the economic, physical and social well-being of their hosts.'

The last word of this definition, 'host', implies an invitation from people who are happy to receive tourists. This term has received increasing criticism from academics, NGOs and the more socially aware quarters of the tourism industry, as levels of cultural and environmental awareness have grown since the early 1980s. It is now recognised that in some cases tourism is something that is tolerated or even forced upon communities as opposed to being welcomed.

Besides referring to the impacts of tourism, Mathieson and Wall's definition adds a further dimension to the concept of tourism by introducing a behavioural dimension, that is, the 'study of people away from their usual habitat'. Given that tourism would not exist without tourists, understanding the motivations of tourists and the effect of their behaviour on the environments of destinations, are areas of interest to social psychologists, sociologists and anthropologists. In a later definition of tourism, Bull (1991: 1) reiterates the behavioural and impact aspects of tourism, whilst also introducing a resource dimension: 'It [tourism] is a human activity which encompasses human behaviour, use of resources, and interaction with other people, economies and environments.'

From Bull's definition tourism can be interpreted as a form of development involving the use of natural resources. This adds another perspective to tourism's interaction with the environment, as a user of natural resources for wealth creation. The idea of wealth creation through the use of the natural and cultural environments for tourism has received increased international attention since the success of General Franco's policies on tourism development in Spain in the 1950s, as is discussed later in this chapter. Today, the focus of using tourism for national wealth creation lies outside Europe, predominantly in the countries of the less developed world. However, the extent to which environmental resources should be used for tourism is both debatable and contentious, raising ethical and political questions. Similarly, ethical questions can be raised over tourism's 'interaction with other people', and the extent to which this is beneficial for local or indigenous communities. These are themes which are developed in the course of this book.

Although there is no definitive definition of tourism, this brief introduction to tourism demonstrates its complexity, and that it is about much more than 'going on holiday'. Tourism is based upon the economic and social processes that are occurring in the environments of the societies where tourists originate from. Its development in destinations involves the use of physical and natural resources and will subsequently impact upon the economies, cultures and ecology of the destinations it develops in.

Is there a 'tourism industry'?

According to the World Travel and Tourism Council (WTTC) (1999), travel and tourism contributed directly and indirectly to the global economy in 1999:

- 11 per cent of Gross Domestic Product;
- 200 million jobs;
- 8 per cent of total employment; and
- will generate 5.5 million new jobs per annum until the year 2010.

Although these represent impressive statistics and reference is often made to tourism being one of the world's largest industries, trying to define the 'tourism industry' is actually extremely difficult. The problem of defining what is meant by the term is summarised by Lickorish and Jenkins (1997: 1): 'The problem in describing tourism as an 'industry' is that it does not have the usual production function, nor does it have an output which can physically be measured, unlike agriculture (tonnes of wheat) or beverages (litres of whisky).'

They also add that the vague nature of the tourism industry has made it difficult to evaluate its impact upon the economy relative to other economic sectors. Similarly, in destinations where tourism development has taken place and environmental problems have arisen, it is not always that easy to disaggregate tourism's contribution to these problems from the contributions of other economic sectors.

Murphy (1985) suggests that a tourism industry does not exist because it does not produce a distinct product. He continues to point out certain industries such as transport, accommodation, and entertainment are not exclusively tourism industries, for they sell these services to local residents as well. Another key difference between tourism and other industries, is that it is the consumer who travels to the 'product', and not vice versa. The major inference of this last point is that the physical and cultural characteristics or qualities of destinations can be treated as a form of product, to be sold in the market-place. It is these characteristics of environments that create expectations in tourists and form a vital part of their experience.

Yet within tourism practices are adopted that are familiar to those of industry. For instance a common expression used in connection with tourism is 'mass tourism'. Mass tourism involves tour operators compiling a standardised package, at the very minimum usually comprising of transport and accommodation, which is then sold into the market-place *en masse* to millions of consumers. This 'package holiday', which relies on mass consumption and sales to keep the prices low, displays the characteristics of 'Fordist' production. This is a term used to describe the mass conveyor belt techniques pioneered by Henry Ford in the motor car industry in the early twentieth century, which paved the way for mass car ownership, by keeping production costs low. Although the development of mass tourism has meant that millions of people have had the opportunity to travel to different countries, its development in destinations is often associated with environmental problems, such as pollution and a loss of local culture. Commenting on mass tourism Poon (1993: 4) writes:

> Mimicking mass production in the manufacturing sector, tourism was developed along assembly-line principles: holidays were standardised and inflexible; identical holidays were mass produced; and economy of scale was the driving force of production. Likewise, holidays were consumed *en masse* in a similar, robot-like and routine manner, with a lack of consideration for the norms, culture and environment of host countries visited.

Another comparison between tourism and other industries is somewhat polemically given by Krippendorf (1987: 19): 'The timber industry processes timber. The metal industry processes metal. The tourist industry processes tourists.'

Although the existence of a tourism industry is debatable there are definite types of businesses that are specifically orientated to providing the services that meet the needs of tourists. These are:

- travel agents and tour operators;
- airlines; and
- the international accommodation sector.

All these sectors are associated with facilitating travel. Some of these companies operate on a global scale such as Sheraton and Hilton International Hotels, whilst some tour operators have grown into major international businesses and are now listed on the stock exchange, for instance the Thomson Travel Group and Air Tours in the United Kingdom. Similarly, some airlines have become transnational businesses, with companies such as American Airlines and British Airways seeking strategic alliances to increase their global market share.

In summary it is apparent that the 'tourism industry' cannot be considered as being similar to other industries as it does not produce a single identifiable product and neither are many of its services used exclusively by tourists. Essentially when the term 'tourism industry' is used, it is important to recognise that it is an amalgam of different businesses and organisations, connected by the common factor of providing services in some capacity to tourists.

Tourism as a system

Another approach to understanding tourism is to think of tourism as a system, incorporating not only businesses and tourists, but also societies and environments. Some authors interpret the different components of tourism as being interlinked thereby forming a 'tourism system'. For instance Gunn (1994) advocates that tourism should be interpreted as a system, adding that every part of tourism is related to every other part, and that no manager or owner involved in the tourism system has complete control over his or her own destiny. It is therefore important that managers involved in any part of this system understand its complexity and possess an holistic *vis-à-vis* reductionist view of their business operations. The decisions and actions that are taken by businesses will have consequences for other components of the system. For instance, the decision of a tour operator to axe a particular destination from their schedule, will have economic and social consequences for the businesses and local community in that destination who rely upon their trade.

According to Page (1995), the advantage of a systems approach is that it allows the complexity of the real life situation to be accounted for in a simple model, demonstrating the inter-linkages of all the different elements. Mill and Morrison (1992) use the analogy of a spider's web to illustrate the inter-relatedness of different parts of the tourism system, in which touching one part of it induces a ripple effect throughout the web. According to Laws (1991) the advantages of interpreting tourism as a system are that it avoids one-dimensional thinking and facilitates a multi-disciplinary perspective. Such an approach is beneficial with a topic that can be interpreted from a range of disciplinary perspectives including economics, psychology, sociology, anthropology and geography. The components of the tourism system modified from Laws (1991) to include a heightened environmental perspective are shown in Figure 1.1.

This model incorporates a range of different elements which together form the tourism system. Important inputs to the system from an environmental perspective include natural and human resources, the use of which are encouraged by both consumer demand in the market system for tourism, and government policy aimed at increasing entrepreneurial activity and inward investment in the sector. Within the overall system, three distinct subsystems are recognisable, all of which overlap and are interrelated. Incorporated in these subsystems are the businesses that have been developed to cater primarily for tourists, such as tour operators, international hotel companies, global airlines, and locally owned tourism enterprises. Within the destination subsystem, the importance of natural and cultural attractions is emphasised, as the basis for attracting tourists. The outputs of the system, which may alternatively be expressed as outcomes, suggest that tourism will bring environmental and cultural changes. These changes illustrate the dichotomy of tourism in the sense that they can be either positive or negative. Tourism can both conserve and pollute the physical environment, whilst it can also bring positive and negative cultural changes, such as employment opportunities for women or result in women being forced into prostitution. Similarly it can create economic opportunities for communities but also can result in an economic overdependence on tourism and encourage price inflation. Another output of the system, which is essential for ensuring the profits of enterprises based upon tourism and helping to secure the economic benefits desired by governments, is tourist satisfaction.

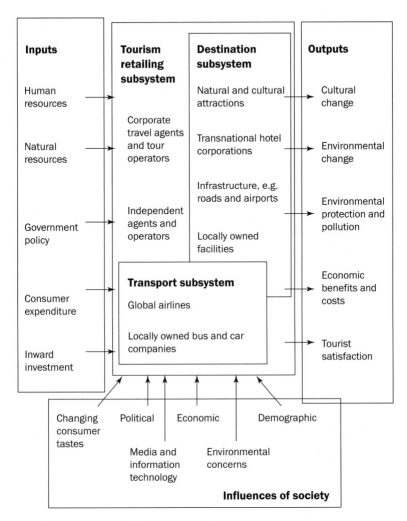

Figure 1.1 *The tourism system: an environmental perspective*
Source: After Laws (1991)

Finally, the tourism system is subjected to a range of influences exerted by changes in society. These may be classified, using a term borrowed from Poon (1993), as 'framing conditions'. Within the context of this model the term applies to those conditions in society which influence the working of the system. For instance, Poon (1993) refers to changing consumer tastes typified by the emergence of the 'new tourist'. These tourists display characteristics of being more environmentally aware, independent, flexible and quality conscious, than the tourists who form the bulk of the mass market. Subsequently the tourism retailing subsystem must adjust its product to facilitate this new market segment, and local governments and municipalities must plan and develop their destination in a way to attract this market segment. Economic, technological and political changes can also influence tourism by making it accessible to a wider proportion of the population. For instance, rising levels of disposable income, longer holidays, and greater political freedoms will all encourage travel. Similarly, technological advancement, such as the development of the jet engine, has made international travel easier and encouraged tourism.

> **THINK POINT**
>
> Why is it more appropriate to think of tourism as a 'system' rather than as an 'industry'?

Media and information technology developments have made more information and images available about potential tourism destinations than in any previous period in the history of society. Increasing access to, and use of, computers, means it is also possible to book airline seats and holidays directly from home via the computer terminal, facilitating travel. At the end of the twentieth century environmental concerns began to exert an influence upon tourism, as suggested by the emergence of 'new' forms of tourism, such as 'ecotourism', and evidence of a growing environmental commitment in some quarters of the tourism industry, themes which are explored later in the book.

The growth in demand for tourism

Although tourism is a familiar aspect of contemporary life, particularly for the majority of people living in the countries of the developed world, it is only relatively recently that it has emerged as a significant aspect of society. The growth in demand for tourism is a reflection of a range of changes that have occurred in society, particularly since the onset of the Industrial Revolution. This section of the chapter examines the history of tourism and reasons to explain its growth, and continues to examine a type of tourism which has been particularly popular in the latter half of the twentieth century, international mass tourism. The significance of the mass participation in international tourism from an environmental viewpoint, is that an increasing number and variety of physical and cultural environments are being exposed to tourism, with a range of consequences.

Pre-industrial tourism

Tourism is not something that happens by chance but is an activity that has developed as a consequence of the type of societies in which we live. Certainly the tendency for people to live in urban areas would seem to increase the propensity for tourism. Although we tend to think of mega-cities as very much a phenomenon of contemporary times, large cities have existed in ancient history, for example Carthage at its fall in 146 BC had a population of 700,000 and Augustan Rome a population of 1 million (Goudie and Viles, 1997). The desire of the Romans to escape the heat of Rome in summer time led them to travel to seaside and mountain villas (Holloway, 1998). Even in Roman times there was evidence of the development of a hierarchy of resorts, possessing distinct types of cultural environments, and attracting different market segments. Holloway (1998: 17) comments:

> Naples itself attracted the retired and intellectuals, Cumae became the resort of high fashion, Puteoli attracted the more staid tourist, while Baiae, which was both a spa town and a seaside resort, attracted the down-market tourist, becoming noted for its rowdiness, drunkenness and all-night singing.

Whilst travelling to resort areas, wealthy Romans would rest at their own private villas *en route* which although only used for three or four nights per annum were fully staffed by servants and elegantly furnished, whilst commoners rested in tavernas built by farmers on the highway between Rome and the coast (Eadington and Smith, 1992). Long-distance travel was also facilitated in the Roman Empire by the development of a transport infrastructure, the need to use only one currency whilst travelling the Empire's length from Syria in the east to Hadrian's Wall in Britain in the west, and the requirement to speak only Latin.

However, until the onset of the Industrial Revolution in the second half of the nineteenth century, the role of the environment in travel was not usually associated with aesthetic pleasure but with the benefits it could offer an individual in terms of health, religious worship, or education. Travel for healing and religious purposes, can be dated to pre-Roman times to Ancient Greece, with pilgrims travelling to visit the sites of healing gods. Religious pilgrimages have up until the last two hundred years, generally been undertaken in pretty arduous conditions, and the willingness to travel as a pilgrim was seen as a measure of a person's devotion to their god.

After the collapse of the Roman Empire in the west in the fifth century AD and the onset of the Middle Ages, travel became more difficult, and from the evidence of historical records was very limited. Travel was arduous, mostly undertaken out of a necessity to trade, or as previously stated to prove one's religious devotion through pilgrimage. However, the early seventeenth century saw the appearance of the 'Grand Tour', a direct outcome of the freedom and quest for learning heralded by the Renaissance, a period marked by a rediscovery of the classical teachings of the civilisations of Rome and Greece. Holloway (1998) links the establishment of the Grand Tour to the reign of Elizabeth I, when young men seeking positions at court were encouraged to travel to the Continent to finish their education. Essentially the Grand Tour consisted of a tour of European culture for aristocratic young men, the absence of references to women in the accounts of the Grand Tour is noticeable, and the aristocracies of Britain, France, Germany and Russia were particularly involved in it. The influence of the Grand Tour on subsequent attitudes to travel was notable, as Towner (1996: 96) comments: 'The Grand Tour, that circuit of western Europe undertaken by the wealthy in society for culture, education, health and pleasure, is one of the most celebrated episodes in the history of tourism.'

For the first time since Roman times, foreign environments were seen as being pleasurable, stimulating and educative. The Grand Tour included visiting the major cultural centres of Europe, and lasted from the beginning of the seventeenth century through to the onset of the Napoleonic Wars, in the first part of the nineteenth century. The tour lasted an average of three years and the gentleman would be accompanied by a private tutor. According to Gill (1967) one of the most notable tutors was Adam Smith, the eminent economist and the author of the seminal *An Inquiry into the Nature and Causes of the Wealth of Nations* (1776), in which he advocates trade liberalisation and free market economics. Although the Grand Tour is predominantly viewed as an aristocratic activity, Towner (1996) suggests that this impression is probably a consequence of the likelihood that the written records of the most prominent people are the most probable ones to have survived. In Towner's view, social participation was likely to have been wider than just the aristocracy, with a possible figure of 15,000–20,000 British tourists partaking in the Grand Tour when it was at its zenith in the mid-eighteenth century.

The development of a travel culture during this period was supported and aided by the appearance of guide and travel books such as William

Thomas's *The History of Italy* (1549). Health aspects also added another dimension to the Grand Tour, for instance Montpellier in France developed as a place to go to to counteract the effects of consumption, and places in the French Riviera such as Nice also developed as places to go to for health cures. Nash (1979) describes how Nice became established as a destination for health tourism in the eighteenth century. By the nineteenth century there were accounts of the town filling with visitors from England escaping the winter, although as Nash points out, by this stage healthy tourists probably outnumbered the infirm. It is interesting to reflect how the purpose of tourism affects the seasonal pattern of arrivals. Today, in contrast to the earlier winter visitors Nice experiences its high season in summer, as tourists arrive primarily for recreational purposes in contrast to the earlier health-driven winter visitors.

Industrialisation and tourism: laying the conditions for the growth of contemporary tourism

The Industrial Revolution, the origins of which can be traced to the mechanisation of cotton and wool production in the north of England in the last quarter of the eighteenth century, brought great economic and social changes in society. A major change dating to this period was the urbanisation of societies of western Europe and North America. Not only did this have the effect of removing people from direct contact with nature but also led to changes in established community structures. Subsequently, perceptions of what were desirable landscapes began to change, and for some sociologists the city became associated with the isolation of the individual. In the next chapter these themes are discussed within the context of changing perceptions of landscapes and the reasons to explain why people travel.

The Industrial Revolution also made society much more time-conscious than it had been previously. In agricultural society the pattern of labour had been determined by the seasons, now it became highly structured around the need to keep industrial production functioning, with man, woman and child often working a 6-day and 70-hour week (Clarke and Critcher, 1985). Drinking whilst at work and breaking-off from work to attend to domestic affairs, which were not unfamiliar work practices, were aspects of work life that factory owners could not tolerate. Yet the interaction of people when performing tasks and enjoying leisure together, encompassed within the same spatial area which typified work and leisure practices in pre-industrial Britain, were essential for establishing agricultural communities. During the Industrial Revolution, work and leisure became highly differentiated, upon the criteria of time and spatial zones. This pattern is reflected in contemporary tourism, as we take defined periods of time off work, and travel long distances away from our home environment to other destinations.

The realisation that a healthy workforce was more likely to be a productive one, and perhaps the fear of a working-class uprising, led to the passing of the Bank Holidays Act in 1871 in Britain. This was a significant act, in the sense that, for the first time, taking time off work had been formalised by an Act of Parliament. With the evolution of the trade union movement and more socially progressive governments, the numbers of days off work for holidays increased, although it was not until 1938 that the Holiday with Pay Act was passed in the United Kingdom. This Act gave workers the right to one week's paid leave from work (all leave from work before this date had been unpaid).

Another major change associated with the Industrial Revolution was a technological advance in transport. Key technical advancements that encouraged mass participation in tourism in the nineteenth century were the development of the railways and steamships. Until the nineteenth century travel was largely dependent upon the horse and carriage. Such journeys were arduous, for example a journey of approximately 640 kilometres between London and Edinburgh took 10 days by horse and carriage (Holloway, 1998). In the latter part of the nineteenth century, the development of railroad systems and passenger steamship services made travel considerably easier. This advanced technology also required a much higher level of investment than the stage coaches that had been the primary means of transport before then, subsequently necessitating the sale of large quantities of passages to secure a return upon one's investment. By 1904, P&O passenger services had already begun cruising, taking first class passengers to Australia and India (Page, 1999).

In the twentieth century, widespread ownership of the motor car and the development of the jet engine in the 1950s, have facilitated travel by making it easier and quicker. The car has given people greater control over their own travel arrangements than at any time in the past, and the provision by governments of a good infrastructure of roads in many countries has meant that going on holiday in the motor car has become a popular option. The development of jet-powered aircraft has led to travelling times between countries being dramatically reduced, facilitating the access of tourists to foreign countries and increasing the propensity for international travel.

The development of the railways, steamships, motor car and jet engine were important technical changes that encouraged travel. Not only did they make travel quicker but it became safer and more comfortable. The high financial investment in the railways, steamships and aircraft, also made it necessary for operators to encourage as many passengers as possible to use their services, encouraging mass participation in travel. In more contemporary times, the effects of pollution associated with travel for tourism is a major concern, as is discussed in Chapter 3.

In current society, the role of information technology is an increasingly important influence upon tourism. There is little doubt that we are living at the time of an information technology revolution, which at the start of the twenty-first century is still only beginning. Just as the development of photography and cinema offered images of foreign lands to a widening audience at the beginning of the twentieth century, the development of an increasingly sophisticated media, involving satellite and cable television, does so today. Computer technology also increasingly plays a role in many people's lives in developed countries. By the early 1990s, not only was it possible to find out the price, availability and location of a holiday spot, but the technology was already available for clients to take a visual tour of the hotel they would be staying in and the rainforest they would be walking in (Poon, 1993).

Besides technological advancement, the growth of tourism has also been encouraged by the emergence of tour operators. Three of the best-known names in the tour operating and travel agency sector are 'Thomas Cook', 'Lunn Poly', and 'American Express', all of whom developed their tour-operating businesses in the nineteenth century. Thomas Cook organised his first fee-paying trip in 1841 as secretary of the Midland Temperance Association with 570 members travelling from Leicester to Loughborough (Holloway, 1998). As Page (1999) remarks, it is rather ironic that the package tour now thrives on an image of sun, sea, sex and booze. By the 1860s Cook had already developed tours to Europe and America, and in 1869 offered the first escorted tour to the Holy Land (Boorstin, 1961). In his first nine years of business, Thomas Cook handled more than one million customers. By the end of the nineteenth century, Sir Henry Lunn was organising trips to the European Alps for British people interested in winter sports, whilst in America the

American Express Company had begun to offer Americans help in securing railroad tickets and hotel reservations.

The twentieth century saw a reduction in the real price of travel, with economies of scale becoming possible for suppliers, as more people wished to travel. The reduction in the real price of travel is illustrated by trans-Atlantic travel between the United States of America and Britain. At the beginning of the twentieth century to cross the Atlantic by passenger ship was a very rare and special event, restricted to a very few, and subsequently expensive. For instance a first-class ticket on the ill-fated passenger ship the Titanic in 1912, cost US $3000 or approximately US $127,000 in today's money, whilst a third-class ticket cost US $40 or US $1,696 in today's money (Ezard, 1998). By the end of the twentieth century, during which time incomes had risen manyfold compared to those at the beginning of the century, a typical discounted air fare between London and New York cost approximately US $250.

The changing economic and social conditions in society that are associated with the Industrial Revolution led to an increasing participation in tourism by members of different social classes. Two phases of mass participation in tourism are evident, with the first wave having taken place in the nineteenth century, fuelled by a quadrupling of the real national income per head and the development of the railways (Urry, 1990). Primarily domestic tourism, it involved the movement of thousands of working-class people from the city areas to the coast, leading to the growth of seaside resorts as holiday destinations in northern Europe. This trend of domestic tourism continued for the first part of the twentieth century, economic recession in the 1930s combined with two world wars greatly restricting the opportunities for the growth of international tourism. The second phase of mass tourism occurred after the Second World War and this time had an international perspective as opposed to a domestic one.

The development of mass participation in international tourism

The post-Second World War years saw the beginnings of the demise of the northern European seaside resorts, a period marked by the origins of what can be termed the 'second wave of mass tourism', this time international. Mass international tourism began in the 1950s, involving the movement of thousands of tourists from the United Kingdom to

Spain, ultimately leading to the development of the western Mediterranean coastline for tourism. A range of factors amalgamated to encourage the movement of tourists to Spain, including the following: an increasing level of disposable income from the late 1950s; a surplus of Second World War aircraft which could be used to provide cheap transport from the UK; the development of tour operators in the United Kingdom who promoted the image of Spain; the encouragement of tourism development by the Spanish dictator General Franco; and the availability of cheap land in Spain for hotel development. By 1955 two million people from the UK were travelling abroad (Page, 1999). The subsequent availability of cheap package holidays to Spain brought foreign travel within the reach of working-class people for the first time, the vast majority having previously had no opportunity for international travel, beyond the experiences of working-class men as soldiers in the Second World War. The factors leading to the growth of Spain as a tourism destination are discussed more fully in the case study in Box 1.1.

Box 1.1

Case study: the development of international tourism to Spain

Although Spain is the country that is probably associated most closely with international mass tourism, its development as a destination for millions of tourists is recent. Spain was never an established part of the Grand Tour, its landscape and culture not regarded as being particularly attractive, and its geographical position rendered it fairly inaccessible. It was not until towards the end of the nineteenth century, with the development of the Romantic movement in northern Europe, that the wild landscape of Spain combined with its medieval and Moorish culture began to be considered attractive by an elite group of travellers, and Spain began to develop an image as being exotic. Its geographical remoteness combined with a lack of infrastructure development, the civil war of the 1930s and two world wars in the first half of the twentieth century, meant Spain retained an exotic image even until the 1950s.

In the 1950s, today's heavily developed tourism areas of the Costa Brava and Costa del Sol began to receive foreign visitors. So rare were foreign visitors in post-war Spain, their movements were monitored by the Civil Guard, a legacy of foreigners who fought in the Spanish Civil War on the side of the Republicans. For example, in 1955 Torremolinos (a name now synonymous with large-scale tourism development) was a poverty stricken

continued

fishing village, where villagers grafted a hard living from the land and sea. The first foreign visitors to arrive in Torremolinos were wealthy foreigners who did not want to walk in the Swiss Alps or sojourn on the French Riviera. The style of the first hotels reflected a luxurious peasant style and by the 1960s Torremolinos had become a highly fashionable resort. Beachfront property increased in price twentyfold in two years, and a villa worth £1,000 in 1955, was sold as a site for a hotel in 1963 for £146,000. Torremolinos's popularity meant increasing numbers of visitors arrived on shuttle buses from expanding airports, which led to increased hotel development. Ultimately the rich and the artistic types deserted Torremolinos and went to nearby enclaves at Marbella or as far afield as Bali and Morocco.

Apart from illustrating a typical cycle of resort development, with a destination area being discovered by a wealthy elite leading to rapid unplanned tourism growth determined by market forces, Torremolinos also illustrates the economic opportunities to be gained from tourism. General Franco encouraged the growth of tourism, based upon a need to gain political recognition of a regime that was viewed with international suspicion, and also as a means of attracting foreign exchange and modernising other sectors of the Spanish economy. Spain was marketed as an exotic, low-cost holiday destination, distinct from its chief catchment area of northern Europe. The growth of tourism to Spain was also fuelled by the development of the jet engine, bringing Spain closer to countries of northern Europe in terms of travelling time, and the expansion of the tour-operating industry particularly in the United Kingdom. Subsequently, Spanish tourism has undergone a major expansion, international arrivals increasing from 700,000 visitors in the early 1950s, to 4 million by 1959, 40 million by the early 1980s, to 47 million in 1998.

Sources: Moynahan (1985); Pi-Sunyer (1996); Barke and Towner (1996), World Tourism Organisation (1999)

The success of Spain in using tourism as a catalyst for economic growth and development, has encouraged governments of other countries to develop tourism as a part of economic policy. Governments can offer a range of financial incentives to encourage tourism development, including the giving of grants and loans at beneficial interest rates for tourism development, relaxation of import duties and tax breaks. Examples of actual incentives that have been used to encourage tourism development are shown in Box 1.2.

The combination of changing social conditions in society and the encouragement of tourism development by national governments, has led to a rapid increase in the number of people travelling internationally post-1950, a rate of increase that is projected to continue until at least the year

---

Box 1.2

Examples of investment incentives used by governments to encourage tourism development

Greece: Incentives were made available for the country's economic and regional development through Law 1262 passed in 1983. Tourism is considered a 'productive investment' and incentives are available for the construction, extension and modernisation of hotels (up to 300 beds), winter sports facilities, spas, tourist apartments, and for the renovation of traditional houses into hostels or hotels. Typical financial incentives include grants, preferential grants, interest rate subsidies, loans, fiscal incentives, and tax allowances.

Malaysia: Reliance is placed upon fiscal incentives rather than grants or loans. Companies may be given 'tax holidays' of up to five years providing the development meets certain criteria, such as the number of employees generated, and the geographical locality of the investment. Tax credit of up to 100 per cent is also given on capital expenditure incurred during the first five years of the operation of a tourism facility.

Source: Bodlender and Ward (1987)

2020. The actual recorded number of international arrivals in 1950 was 25 million, in 1980 this had risen to 270 million, by 1998 it had reached 600 million, and by 2020 it is projected to be 1602 million (World Tourism Organisation, 1998a). This trend in international arrivals is displayed in Figure 1.2.

The dramatic growth in tourism is further underlined when the decade upon decade increases are calculated since 1950. Average increases for each decade are over 50 per cent as portrayed in Figure 1.3.

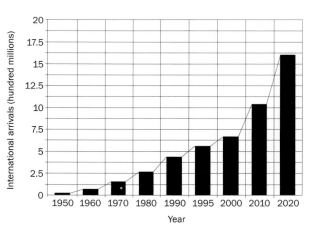

Figure 1.2 *International arrivals: actual figures, 1950–2020*

Source: World Tourism Organisation (1998a)

This spatial spread of international tourism

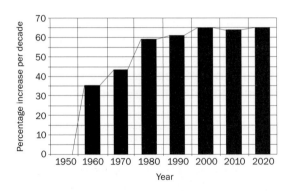

Figure 1.3 *International arrivals: percentage increase per decade, 1950–2020*

Source: World Tourism Organisation (1998a)

demonstrates a pattern of 'core and periphery' relationships. This was a term that originated from the development sociologist Gunther Frank, to explain the global pattern of economic dependency of lesser developed countries (the 'South') upon the more developed countries of Europe and USA (the 'North'), which according to Frank can be traced to historic patterns of colonial imperialism. The same analogy of core to periphery relationships can also be used in tourism to describe the dependency of many less developed countries (LDCs) upon the markets of countries of Europe and the USA for international tourism. As Cohen (1995: 13) remarks, 'Tourism is essentially a modern western phenomenon', although he adds that modern tourist travel has spread in recent decades to contemporary Japan, Taiwan, and Korea. Regardless of nationality and culture, it would seem tourism is capable of being sold into any strata of society, which has both time and money. The term also refers to the dependency of many developing countries upon transnational corporations, such as the Sheraton Group, for foreign investment to develop their tourism sector. Additionally, LDCs may be reliant upon loans, from foreign governments or banks, to develop tourism. This has led many critics of tourism to complain that tourism represents little more than a continuing form of post-colonialism, where all the power and control in the global tourism system remains in the 'North', and within which the environments and cultures of the 'South' are exploited.

Although the figures of expenditure on international tourism support the theory of the dominance of the traditional core areas of Europe and the USA, it is evident that new cores are emerging. In 1997, the world's top six spending countries on international tourism were the United States of America, Germany, Japan, the United Kingdom, Italy and France (World Tourism Organisation, 1999). However, the complete list of countries, ranked by order of consumer spending on international tourism, reveals some interesting changes in the world order. Apart from Japan in third place, in tenth place is China, the Russian Federation is in eleventh place, and Taiwan and the Korean Republic are in seventeenth and eighteenth place respectively, ahead of Australia, Norway, Spain and Denmark (World Tourism Organisation, 1999). The list lends support to the theory that economic development and political freedom will lead to an increased demand for tourism, and also goes some way to supporting the idea that if global political influence and economic power in the nineteenth century rested with Europe and in the twentieth century with the USA, then the twenty-first century is likely to belong to Asia.

Many countries of the world, which had previously experienced little tourism, developed as international destinations in the last four decades of the twentieth century. Areas of the world and countries as diverse as Greece, Turkey and other parts of the eastern Mediterranean; the Caribbean; Thailand and other countries of south-east Asia; China; parts of Africa; South America; and the Pacific Isles have all developed large tourist industries based upon the attraction of their cultural and physical environments. One of the most notable emerging tourism destinations is China, which received 24 million tourists in 1998, making it the fifth highest ranked country expressed in terms of incoming tourism.

At the beginning of the twenty-first century international tourism has expanded to include a variety of destinations and different environments. As Poon (1993) points out, tourists are demanding the experience of new cultures, physical environments and activities. This has led to indigenous cultures and special physical environments such as rainforests, coral reefs and polar areas becoming the focus of tourism. With the technological advancement of the twentieth century it would seem that no environment is too remote or inaccessible for tourism. The peripheries of tourism continue to be stretched further afield and the next periphery of tourism is likely to be space. Already one American company 'Zeaghram Space Voyages' has sold more than 250 places at £60,000 each to visit space, and plans are already developed for the construction of a space resort having room for approximately 100 guests by the year 2017 (TTG, 1999).

> **THINK POINT**
>
> Describe the changes that have occurred in the environments of societies where tourists originate from which help to explain the growth in demand for international tourism.

Summary

- Demand for tourism is not something that occurs by chance but is symptomatic of changes in the environment of societies that tourists originate from. Economic and social changes associated with the Industrial Revolution have had a major effect in shaping patterns of contemporary tourism.

- Tourism is a highly complicated amalgam of different parts. These include the environments of where people come from and where they go to. It is useful to think of tourism as a system, including a range of different inputs such as government policy, entrepreneurial activity, and human and natural resources. Critical to the system is the tourism industry which 'processes' these inputs to meet the needs of the tourists. Outputs of the system include economic opportunities and changes to the cultural and physical environments of destinations. All the time the system is subject to a range of external forces existing in society.

- There are now over 650 million international tourist arrivals in the world and the number is expected to be over 1,600 million by the year 2020. Tourism is encompassing an increasing range of physical environments and indigenous cultures as tourists demand new experiences. In turn this means a growing number of environments are being changed by tourism.

Further reading

Holloway, C. (1998) *The Business of Tourism*, 5th edition, Harlow: Addison Wesley Longman.

Towner, J. (1996) *An Historical Geography of Recreation and Tourism in the Western World: 1540–1940*, Chichester: Wiley.

Urry, J. (1990) *The Tourist Gaze: Leisure and Travel in Contemporary Societies*, London: Sage.

2 **Perceptions of environments for tourism and ethical issues**

- The meaning of environment
- Changing perceptions of the environment
- Tourist motivations and types of tourists
- Ethical considerations of tourism

Introduction

The environment is often referred to as the key component of tourism. Chapter 1 introduced the idea of viewing tourism as a system, linking the environments where tourists go to with the society they come from. This chapter examines the meaning of the term 'environment', how tourists perceive and interact with different environments, and raises ethical considerations of how the environment is used for tourism.

What is meant by 'environment'?

One of the two key words of the title of this book was examined in Chapter 1, and now the question is asked, what is meant by the other key word, 'environment'? It is a term that is used in everyday conversation but its actual meaning and relationship to tourism is perhaps uncertain. According to Allaby (1994: 138) the environment is:

> The complete range of external conditions, physical and biological, in which an organism lives. Environment includes social, cultural and (for humans) economic and political considerations, as well as the more usually understood features such as soil, climate, and food supply.

Similarly, Collin (1995: 83) defines the environment as:

> The environment is anything outside an organism in which the
> organism lives. It can be a geographical region, a certain climatic
> condition, the pollutants or the noise which surrounds an organism.
> Man's environment will include a country or region or town or house
> or room in which he lives; a parasite's environment will include the
> body of the host; a plant's environment will include a type of soil at a
> certain altitude.

From these definitions it is evident that the environment of tourism can
be viewed as possessing social, cultural, economic and political
dimensions, besides a physical one. Given that both definitions specify
the environment as the external conditions in which an organism lives,
the demand for tourism can be interpreted as being a consequence of the
interaction between the social, cultural and economic environments
where tourists originate from.

To make the concept of the environment a little less abstract, it can be
classified into natural, built (or human-made) and cultural types (Hunter
and Green, 1995). These three types are not exclusive and can be viewed
as being interrelated by human influence. Evidently the built and cultural
environments are directly reflective of aspects of human behaviour. The
actions of humans have also historically altered the natural environment,
through activities such as farming, industralisation and urbanisation.
Interference with the natural environment is now being experienced on a
global level in regions where humans do not live. For instance, pollutants
that are produced in industrialised societies, such as DDT, lead and
sulphates, have all been found at the North and South Poles (Goudie and
Viles, 1997). Within the context of tourism, it is the characteristics of the
natural, cultural and built environments of destinations that pull tourists
towards them. All environments are different and have their own unique
features, and what makes a particular type of environment attractive for
tourism is a function of value judgements and fashions that exist in
society, as the next section of this chapter continues to explain.

Changing perceptions of landscape

A key part of desiring to visit a particular place is a favourable perception
of the environment that lies at the other end. All of us have ideas of what
we may regard as desirable places to visit. However, our impressions of
beautiful landscapes and 'exotic' cultures are subject to modification as

society and fashion changes. As Urry (1995) suggests, a person's wish to visit a particular environment is something that is socially constructed, and depends upon developing a 'cultural desire' for a particular landscape.

Shifts in perception of what are regarded as being desirable landscapes are therefore associated with social and cultural changes in the society that tourists originate from. For instance, a notable change in the itinerary of places to visit within the Grand Tour began in the mid-eighteenth century, as the desire to view picturesque and romantic landscapes became stronger (Towner, 1996). This wish to view picturesque and romantic scenery noted a marked shift from what had previously been regarded as being desirable landscapes. The previous landscapes of fashion were those of the European low countries, that is, Belgium and Holland, because they illustrated the human ability to control and dominate nature to provide agriculturally productive terrain. This view of desirable nature as being agriculturally productive is representative of the separation of landscapes into controlled and wild areas. According to Short (1991), the concept of 'wilderness' developed approximately 10,000 years ago, with the onset of the agricultural revolution and a move to a more settled pattern of living. Previously in nomadic hunter and gatherer societies, no distinction was made between wilderness and the environment.

Two main perspectives on the meaning of 'wilderness' can be recognised. The first is a 'classical perspective', in which the view is taken that the creation of livable and usable spaces, such as urban areas, is a mark of civilisation and progress. The second approach is the 'romantic', in which untouched spaces have the greatest value, and wilderness assumes a deep spiritual significance. The shift to a desire to visit 'wildscape' in the mid-eighteenth century was marked by a preference for the raw power of nature, as manifested in mountains, gorges, waterfalls and forests. Until this time barren and mountainous landscapes had been largely detested and even feared by the majority of the population. For example, Smout (1990) points out that up to the eighteenth century the environment of the Scottish Highlands in Britain had been largely seen as an inhospitable place to think of visiting, and any use of its environment had strictly been in utilitarian terms, such as for the cutting of wood and mining of ores. Urry (1995: 213) adds: 'It is only in the last century [the nineteenth century] that a traveller passing through the Alps would have the carriage blinds lowered to make sure they were not unduly offended by the site.'

Urry (1990) associates the shift in emphasis towards wilder landscape with the development of the 'Romantic movement', which emphasised the feelings of emotion, joy, freedom and beauty that could be gained through visiting 'untamed' landscapes. The Romantic movement was a collection of literary, artistic and musical figures, such as Rousseau, Coleridge, Wordsworth, Chopin, Liszt and Brahms, who highlighted the importance of emotional experiences and having feelings about the natural and supernatural world. The movement was in part a reaction to the scientific thinking of the Enlightenment period, and also to the growing urbanisation of Britain and western Europe, associated with the Industrial Revolution. Indeed, the Romantics represented a form of political opposition to what they perceived to be a loss of community, associated with the migration of people to urban areas from rural environments.

One of the most poignant examples of the work of Romantic literature is the poem, 'The Daffodils', written by William Wordsworth at the beginning of the nineteenth century. The opening verse of the poem is as follows:

> I wander'd lonely as a cloud
> That floats on high o'er vales and hills,
> When all at once I saw a crowd,
> A host of golden daffodils,
> Beside the lake, beneath the trees
> Fluttering and dancing in the breeze.

This verse emphasises the sense of freedom and wonderment about the physical environment that was an essential ingredient of the Romantic movement. It was written about an area known as the English Lake District which is now a premier area for tourism in the United Kingdom, attracting millions of visitors per annum, many of them visiting Wordsworth's house near Grasmere. Ironically two hundred years later the Romantics themselves have become a tourist attraction.

This change in desire, away from visiting landscapes reflecting human dominance to more wild landscapes, was a notable change in the western cultural perception of landscape, and one that has had a major effect on the patterns of contemporary tourism. For instance, Romanticism not only led to the appreciation of mountain areas, but also to an appreciation of the coastline. A key icon of contemporary tourism, the beach, underwent a major perceptual change dating from the Romantic period. Just as mountain environments had predominantly been viewed as hostile and inhospitable places to visit in the eighteenth century, similarly the beach was also viewed with trepidation. For instance, in Daniel Defoe's early eighteenth-century novel *Robinson Crusoe*, the desert island was not equated with escapism and paradise as it is now, but viewed as an environment that was symbolic with abandonment and desolation. To the European explorers of the 'New World', beaches were points of initial contact with foreign culture and potential battlegrounds, as Lěncek and Bosker (1998: xxi) comment:

> They were anxiety-ridden strips of no-man's-land where, according to the journals of Columbus, Cortés, Cook, and Bougainville, Europeans first set eyes on others who, though like them, were yet utterly alien. Here, duels to the death were waged between races and cultures.

Even in more contemporary times, notably the Normandy landings in France in the Second World War, the beach regained its aura of hostility. It is difficult to give an exact date of when the beach became a more popular place to visit. Certainly an association of the coast with health in the nineteenth century, the drinking of sea-water being viewed as medicinal, helps to explain a changing perception of the beach in this period. The process of urbanisation associated with the Industrial Revolution also helps to explain its increasing popularity during the same period as a place to escape to. The natural environment of the beach offered a direct contrast to the squalid urban conditions that existed in many urban areas, offering a sense of both naturalness and timelessness. An example of the squalid conditions experienced by many urban working-class people is given by Engels in his description of life in Manchester in England:

> In a rather deep hole, in a curve of the Medlock and surrounded on all four sides by tall factories and high embankments, covered with buildings, stand two groups of about two hundred cottages built chiefly back to back, in which live about four thousand human beings, most of them Irish. The cottages are old, dirty, and of the smallest sort, the streets uneven, fallen into ruts and in part without drains or pavement; masses of refuse, offal and sickening filth lie among standing pools in all directions; the atmosphere is poisoned by the effluvia from these, and laden and darkened by the smoke of a dozen tall factory chimneys.
>
> Frederick Engels, cited in Foster (1994: 58)

In the latter part of the nineteenth century, the development of purpose-built attractions also aided the evolution of seaside resorts into major tourism destinations. The focus of beach tourism also began to change in this period, from health to pleasure, a change aptly described by Urry (1990: 31): 'In the mid-nineteenth century this medicalised beach was replaced by a pleasure beach, which Shields characterises as a liminal zone, a built-in escape from the patterns and rhythms of everyday life.'

Today the beach represents the main focus of global holiday tourism, being popularised in contemporary film and television, through programmes such as 'Baywatch'. The beach has become an icon of 'dreams', a 'paradise' consisting of an unspoilt environment with beautiful people where everyone enjoys themselves.The popularity of the beach as the primary focus for recreation and tourism is supported by empirical research based upon 542 households in England and Wales, as discussed in Box 2.1.

Box 2.1

The popularity of the beach

In a survey of 542 households in England and Wales, Tunstall and Penning-Rowsell (1998) found that the beach was rated a more enjoyable place to visit than national parks, lakes, rivers, woods, museums, leisure centres or theme parks. People were found to be most attracted by those areas of coastline that were perceived as being undeveloped, possessing characteristics of natural settings such as dramatic views and scenery, and offering peace and quietness. The research also discovered that one of the most important functions of the beach was its ability to reconnect people with their past, when visits were made in childhood and adolescence, thereby permitting an element of continuity and timelessness in a rapidly evolving and changing society. Tunstall and Penning-Rowsell liken this continuity to rereading a favourite book or the family photograph album. The beach also provided an environment for family interaction, with two generations of a family often visiting the beach, and was also found to act as a stress reliever. The ability to be in contact with nature through visiting the beach is emphasised in the following passage:

> Perhaps one of the most important elements in the beach experience is, therefore, the opportunity it gives for tactile close up contact with the natural physical world. . . . There are very few natural environments where children and even adults are allowed, indeed encouraged, to poke about, pick up, touch, shape and play with its physical material and the creatures it supports – crabs, shellfish and worms.

(Tunstall and Penning-Rowsell, 1998: 329)

The desire of the public for the beach to act as a point of contact with nature and the past is emphasised by Tunstall and Penning-Rowsell (1998: 330): 'Thus, the general view appears to be that the seaside should be like the seaside always was.'

Although the phrase as the 'seaside always was', will have varying interpretations for different people, the empirical research supports the popularity of wildscape in society. Such research is also invaluable in giving guidance to planners and developers about the appropriateness of different types of coastal development for recreational use.

Source: Tunstall and Penning-Rowsell (1998)

Promoting the environment as an attraction

Changing perceptions of landscapes, combined with the changing social and economic conditions of the nineteenth century, presented opportunities for entrepreneurs to begin to promote images of the environment to the public to encourage them to travel. Notably, the development of the railways combined with the influence of tour operators such as Thomas Cook and Sir Henry Lunn, meant that by the beginning of the twentieth century images of 'wild' landscapes and foreign cultures had become central to attracting people to travel. Two examples of the use of wildscape to attract tourists are displayed in Figure 2.1 and Figure 2.2.

Although it is not possible to tell from the black and white image reproduced in this book, the original poster shown in Figure 2.1 uses only two colours, black and yellow to demonstrate the austere image of the cliffs and sea. This image of Carrick-a-Rede on the coast of County Antrim in Ireland conveys a sense of wonderment, risk and adventure. It also emphasises, through the use of the imagery of a barely discernible single figure on the suspension bridge, the power of nature and the solitary experience wilderness can provide. A further interesting aspect of the poster which dates to 1904, is that it was produced for the Midland Railways in association with Thomas Cook the tour operator, emphasising the role of the railways and tour operators in developing tourism in this period.

In Figure 2.2, the image on the poster dating to 1908 presents a more 'exotic' kind of travel to the western eye. Again the poster demonstrates the importance of the railway companies in promoting and facilitating travel at this time. It also underlines the post-colonial nature of

Figure 2.1 *'Midland railway: Cook's excursions to Ireland'*

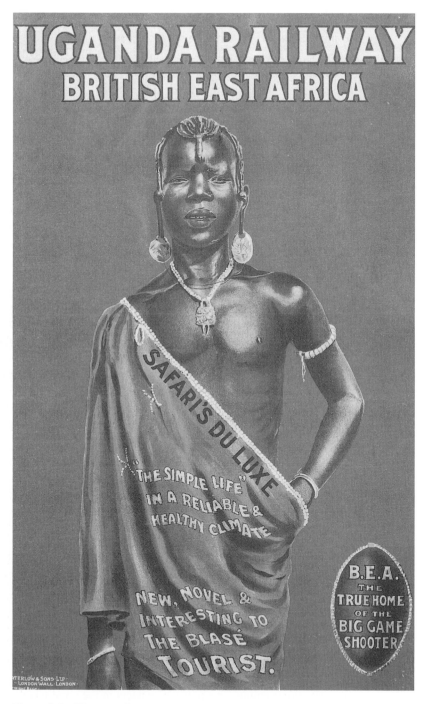

Figure 2.2 *'Uganda railway: British East Africa'*

contemporary tourism, with the countries of Kenya and Tanzania which formed 'British East Africa', now representing two of the biggest receiving countries of British outbound tourists in Africa. The poster also raises ethical issues about tourism and the 'use' of indigenous cultures by the tourism industry for promotional purposes. Clearly, the exotic figure is presented to appeal, as in the wordage on the poster, 'new, novel and interesting to the blasé tourist'. The notion of a 'blasé tourist' at the beginning of the twentieth century may seem surprising, but it must be remembered that this kind of holiday was targeted at an elite in British society, hence the emphasis on 'safari's du luxe'. For this travelling elite, perhaps already overfamiliar with tourism to Europe, new environments were being sought. The other wordage on the poster is also revealing, 'The simple life in a reliable and healthy climate', emphasising the desire for escapism from what was becoming an increasingly urbanised and complex environment.

However, the biggest attraction of travel to 'British East Africa' at this time, as promoted in the poster was the activity of big game shooting. The growth in the popularity of hunting in East Africa is associated with the arrival of white settlers in the nineteenth century. In Europe, hunting was an activity exclusive to the aristocracy, helping to differentiate them from other classes. The settlers now had an enormous shooting estate of their own and the guns proclaimed them as the new aristocrats. Within a few decades of the Europeans arriving the blaubok and quagga were eliminated, both of whom had survived 3 million years of contact with the indigenous people, and men boasted of killing 200 elephants on safari. One notable example of the carnage caused through hunting was an expedition led by Theodore Roosevelt and his son, in which 5,000 animals of 70 different species were killed, including nine of East Africa's remaining white rhinos (Monbiot, 1995).

Although the imagery used by today's tour operators is more subtle than that in Figure 2.2, the messages conveyed in destination advertising continue to rely upon the differentiation of the destination environment from the one at home. This is aptly demonstrated in Figure 2.3, showing a typical image used by tour operators to sell long-haul travel to destinations outside Europe and America. This particular image is of one of the coral islands of the Republic of Kiribati in the Pacific, the first land area to welcome in the new millennium, and ironically one of the areas most under threat in this century from increasing sea levels resulting from global warming. The physical aspects of the coral island, such as the palm trees and white beach, combined with the crystal clear sea, are

Figure 2.3 *Image of Kiribati*

often utilised in promotional images by the tourism industry to attract tourists.

The influence of the 'Romantic' movement on contemporary copywriting in tourist promotion is evident in the following advertisement for the Azores. Under the title 'Europe's last frontier', the Azores are described as:

> There's a place where beauty is not near extinction.
> Where nature and man have found perfect harmony. Where peace is unbroken.
> There's a place where you may still watch whales swimming freely in their natural habitat.
> Where vulcanic heat allows for an exquisite and tasteful way of cooking.
> And where memories of Atlantis still linger along the quiet paths
> and the evergreen tracks of the highlands.

Underneath these words is an image of a whale with its tale protruding from the water. It is not a coincidence that the outline shape of the layout of the words of the advertisement resembles the tail of a whale. These words also place a heavy emphasis upon the nostalgia of the way the natural environment was, in which man and nature are in harmony and symbiosis, a world that has changed little since the times of Atlantis.

Empirical research also supports the importance of 'unspoilt' environments being attractive for tourism. During market research with German tourists, when asked about the characteristics of a destination that give it 'quality', the majority of the statements with the highest percentage response rates were found to be environmentally linked as shown in Box 2.2.

Box 2.2

The main characteristics for 'quality' tourism

Statement	Percentage rating it important
The landscape must be beautiful	46
The atmosphere must be relaxed	46
Cleanliness is a matter-of-course	39
The sun must shine	38
The climate must be healthy	32
Good cuisine must play a part	30
Quietness and very little traffic	29
The surroundings must be typical for the country	28
There must be attractive places for excursions	26

Note: The same person could name several characteristics if they thought them important in determining the quality of a destination.

Source: European Tourism Analysis (1993)

These results indicate that the perception of a 'quality' destination is closely associated with factors relating to its physical and cultural characteristics. Critically, in the eyes of the tourist the landscape must be beautiful, a perception which as previously discussed is culturally determined and thus subject to change. The recuperative aspect of tourism is also evident, with emphasis being placed on a relaxed atmosphere, healthy climate, cleanliness and quietness. Cultural aspects of the environment are also stressed, with a wish for good cuisine and surroundings typical of the country. However, whether tourists would wish to be surrounded with the type of poverty seen in some regions of less developed countries is debatable. The probability is that the majority of tourists would prefer surroundings that reflect their own image of a destination in a positive way, rather than being faced with the unpleasant reality of some aspects of the local society.

From the evidence of tourism advertising, it is suggested that images of 'unspoilt' physical and cultural environments are critical for attracting tourists from countries in advanced stages of capitalism to overseas destinations. The way the tourist gains access to sample these unspoilt environments is by purchasing a 'right' of access via the market system. In this sense, tourism can be interpreted as a form of consumerism, centred upon the consumption of experiences provided by foreign environments.

> **THINK POINT**
>
> Think about images of environments that are attractive to you. What makes them attractive? What kinds of associations do you make with them? How do you think your surrounding environment influences these perceptions you hold?

Tourism as a form of consumption

As societies in the West have entered advanced stages of capitalism, 'consumption' has become a central theme of lifestyle. The concept of consumption surpasses the idea of merely buying something to meet basic physiological needs, such as buying a loaf of bread because of hunger. Using social and human psychological perspectives associated with consumer behaviour studies, it is recognised that by purchasing a certain good or service, a range of needs may be met that go beyond our most basic biological requirements. These include social needs (such as feeling part of a group or having a sense of belonging), the need for love, the need for self- and social esteem, and the need for self-development and self-fulfilment. Such ideas are linked closely to the work of Abraham Maslow (1954) in the field of human psychology.

Sociologists emphasise that through the process of consumption it is possible to differentiate oneself from the crowd, and subsequently gain a sense of identity. The idea of identity being linked to consumption was a theme developed in the late nineteenth century by the American sociologist Thorstein Veblen (1899). Veblen's work was based upon observations of the newly emerging wealthy middle class in America who were making considerable amounts of money from trade and manufacturing. In what was a critical account of their lifestyle, Veblen coined the phrase 'conspicuous consumption', to explain how this class used leisure and consumption to differentiate themselves from the rest of society. Through the purchasing of goods and other ornaments and services, they took on a different identity from the rest of the populace, who could not economically afford to purchase these products. By the end of the nineteenth century, tourism also offered a form of consumption, that permitted social differentiation. Within what was a predominantly patriarchal society, by sending one's wife or daughter on holiday from America to Europe, the message was conveyed that one possessed wealth.

At a similar time to Veblen's observations of what he termed the new 'leisure class' in America, in Germany another sociologist Georg Simmel was also observing and writing about the life of Berliners. His central theme was based upon the interaction between the individual consciousness and the modern city. According to Simmel the complexity of urban living, combined with the need to filter out stimulation to a manageable level, leads the individual to an ultimate sense of indifference about the city, whereas at first it had been perceived as exciting and stimulating (Lechte, 1994). Simmel also emphasised the need for an individual to preserve a level of autonomy in urban areas (Bocock, 1993). One way of doing this, as observed by Veblen, is through 'conspicuous consumption'. In this sense it can be said that the process of urbanisation had a major influence in encouraging consumption beyond meeting basic needs, to provide a form of identity.

Using tourism as a means to achieve social differentiation has become increasingly prevalent in western society. Mowforth and Munt (1998) use the term 'habitus', borrowed from the French sociologist Pierre Bourdieu, to suggest that the types of tourism we participate in carry cultural symbols and meanings. Habitus refers to the ability and inclination of individuals and social classes to adopt objects and practices that differentiate them from others in society. Although the ability to participate in international tourism in contemporary society fails to achieve social differentiation as at the time of Veblen, where we are able to go and what we choose to do on holiday carries cultural messages and differentiates us from others in society. For example in British culture, the decision to go on a package tour to Majorca conveys a very different social and lifestyle message to independently visiting Chang Mai in the north of Thailand, or charting a luxury yacht for a boating holiday off the coast of Turkey.

The link between urbanisation, consumption and tourism has been developed by both Boorstin (1961) and MacCannell (1976) to explain why people choose to travel. Boorstin (1961) proposed that tourism represented nothing more than a form of escapism from urban life, serving little purpose beyond mindless enjoyment. Central to this idea is that the tension of urban living creates a state of anomie in the individual. The concept of anomie was developed by the sociologist Emile Durkheim, to describe the individual's feeling of isolation in a society where the norms governing interaction have broken down, and society has transcended into a state of lawlessness and meaninglessness (Dann, 1977; Slattery, 1991). Viewing tourism as a commodity and questioning its degree of meaningfulness, Boorstin (1961: 91) makes the following comments upon a visit to Hyderabad in India:

> Seated beside me on the plane were a tired, elderly American and his wife. He was a real estate broker from Brooklyn. I asked him what was interesting about Hyderabad. He had not the slightest notion. He and his wife were going there because the place was 'in the package'. Their tour agent had guaranteed to include only places that were 'world famous', and so it must be.

In contrast MacCannell (1976) views tourism not as a function of escapism, but as a pilgrimage of modern man, a quest for authenticity which is felt to be lacking in post-modern societies. MacCannell (1992) adds that the desire to get in touch with a way of life that has disappeared with the growth of urban metropolises, also helps to confirm the relative comfort and material well-being of the tourist, *vis-à-vis* the 'primitive'. For example, in MacCannell's latter argument, the purpose of a visit to 'British East Africa' as portrayed in Figure 2.2, would partly be for the tourist to confirm their way of life as being superior to that of the life of the indigenous person. Although the early works of Boorstin and MacCannell are important for providing early sociological perspectives on tourism, they present two extreme views of the influence of society on motivating people to become tourists. Subsequently, Cohen (1979) criticised their work as being overstated, and suggested there was a lack of sufficient empirical evidence to support either claim. It is perhaps best to picture a continuum between the polarised positions of both Boorstin and MacCannell, along which the influence exerted by society upon the individual to become a tourist rests somewhere between the two. However, the observation that urbanisation has encouraged tourism is supported by the results of a survey shown in Box 2.3, concerning how departure rates for summer holidays relate to the population sizes of municipal areas.

Box 2.3

Departure rates for summer holidays in France between 1969 and 1988

% by place of residence	1969	1985	1986	1987	1988
Rural	19	40	38	40	43
Less than 20,000 inhabitants	38	51	46	47	47
20,000 to 100,000 inhabitants	51	60	53	55	60
100,000 + inhabitants (excluding Paris)	56	64	62	60	60
Paris	83	83	79	75	79

Source: Tourism Planning and Research Associates (TPR), 1995

At face value the positive correlation that exists between the increasing size of the urban area and the increasing departure rate for summer holidays lends support to the theory that urban life encourages tourism, whether for escapism or as a quest for something more meaningful. However, as Tourism Planning and Research Associates (1995) point out, other factors favour the participation of urban dwellers over rural dwellers in tourism. These include higher incomes, smaller households with a corresponding greater flexibility for travel, paid holiday leave, and a freedom from being tied to elements of the agricultural cycle such as looking after livestock.

Although the theories of Boorstin (1961) and MacCannell (1976) are useful in initiating the debate about tourism motivation, subsequent work in the fields of social psychology and sociology suggest that tourism motivation is a highly complex phenomenon, satisfying a variety of individual needs. This complexity means that there is a lack of an agreed psychological theory to underpin an approach to understanding tourism motivation. As Pearce (1993b: 114) comments: 'research into why individuals travel has been hampered by the lack of a universally agreed upon conceptualisation of the tourist motivation construct'.

In one of the most significant empirical studies of tourist motivation based upon tourists staying in Barbados, Dann (1977) found that escapism and the search for social status formed the two most important reasons to participate in tourism, which he subsequently termed 'anomie' and 'ego-enhancement'. Behaviour associated with a feeling of anomie manifested itself by tourists placing an emphasis upon social interaction with family and friends. The need for ego-enhancement is, according to Dann, a reflection of people being denied status in their home society, leading to a desire for a higher status away from the home environment. By visiting a country where the standard of living is substantially lower than in the West, a tourist can live in a style of comparative luxury, albeit temporarily. Additionally, by indulging in a type of lifestyle that can not be enjoyed by the majority of local people, the tourist is conferring their status upon others. Additionally according to the concept of 'habitus', by visiting the 'right' type of destination, it is possible to impress one's peer group, thereby raising one's status.

Besides the needs of social bonding and social esteem, other kinds of psychological needs are also important in motivating people to become tourists. Other suggested psychological determinants of tourism are shown in Box 2.4.

Box 2.4

Psychological determinants of demand

- escape from a mundane environment, akin to Boorstin's (1961) theme of escapism
- relaxation, including mental and physical recuperation
- play, a chance to regress to a childhood state which is not available to adults within the constructs of everyday society
- strengthening family bonds
- prestige – akin to Dann's (1977) idea of ego-enhancement
- social interaction
- sexual opportunity
- educational opportunity
- self-fulfilment – akin to MacCannell's (1976) idea of self-discovery

Source: After Ryan (1991)

Empirical research lends support to tourism fulfilling a range of psychological needs. The desire for a change of lifestyle and the stimulus of a new environment is emphasised by Jamrozy and Uysal (1994), who conducted an analysis of the motivations of the German tourist market. They found that escape and a search for novelty and experience were the two major driving forces of tourism motivation. Other motivations included: being with friends and togetherness; adventure and excitement; doing nothing; and prestige. The multivariate nature of the needs that are satisfied through tourism is a theme that has been developed by Pearce (1988), who suggests that people have 'careers' in tourism, just as at work. Just as in a work career, a travel career is both consciously determined and purposeful, although an important difference is that a career in tourism is more likely to be intrinsically motivated than a work career which for the majority of people is likely to be extrinsically motivated. The basis of the 'travel career' is that motivations for participating in tourism are dynamic and will change with age, life cycle stages and the influence of other people. Additionally, past tourism experiences are likely to influence future behaviour.

In summary, the reasons why people choose to travel are complex and multivariate. The influence of one's home society would seem to be critical in shaping the desire to travel but there is no single explanation of why people choose to travel. However, what is evident from the increasing demand for tourism is that growing numbers of people identify tourism as a medium through which to satisfy their wants and desires.

> **THINK POINT**
>
> What were your motivations for taking your last holiday? To what extent do you feel your motivations were influenced by your home environment, including both people and your physical surroundings?

Types of tourists

Given the complexity of what motivates people to become tourists it is unsurprising that tourists choose to visit different types of destinations and display different types of behaviour within the destination environment. Differences in behaviour are compounded by a range of interrelated factors, such as demographics, culture, lifecycle, level of education, and beliefs and attitudes. The fact that tourists are different means that their interaction with the cultural and physical environments of the destinations they visit will vary. The realisation that tourists are not all the same has meant that various attempts to classify tourists into different types have been made. One of the most acknowledged

Box 2.5

A phenomenology of tourist experiences

Mode	Behaviour
Recreational	Emphasis is placed on tourism as a form of entertainment, restoring physical and mental powers, and endowing a sense of well-being. The tourist is thereby refreshed to return to their society where their centre lies
Diversionary	The tourist finds in tourism a diversion or escape from the boredom and meaninglessness of everyday life. This type of person has no centre but neither are they looking for one
The Experiential	The tourist looks for meaning of life in the culture of the 'other', as the tourist has lost their own centre in their home society. Whilst the tourist observes and interacts with other cultures, she/he remains aware of their 'otherness'
The Experimental	This surpasses the experiential mode in the degree of experience of foreign culture. The tourist experiments in different way of living – for example, on a kibbutz, in search of a spiritual centre. However, they remain unsatisfied by the authenticity of any of the cultures they have visited
The Existential	The individual's spiritual centre is now located in another place away from the home environment. For practical reasons, such as work or the family, the person may not be able to relocate physically but will visit whenever possible

Source: After Cohen (1979)

phenomenological attempts to classify tourists' experiences was made by Cohen (1979). He identified five main categories of tourists as shown in Box 2.5.

Cohen's (1979) central theory to explain the different types of tourist experience is that the significance of tourism depends upon the individual's 'total world view'. This means whether the person orientates to a 'centre' which gives spiritual or cultural meaning to their life, and if so, where this 'centre' is located in relation to the society in which they live. The behaviour of a tourist in a destination will therefore be related to where their centre lies. For example the 'Recreational' tourist would be primarily concerned with rest (or 're-creation') and enjoyment rather than seeking a spiritual centre away from home. Interest in and contact with the local culture is likely to be kept to a minimum and this type of tourist is likely to be content with 'pseudo-events' and inauthenticity. They may choose to surround themselves with the paraphernalia of home, such as familiar daily newspapers, food and drink, emphasising a close contact with the home environment. An example of the development of local tourism services to satisfy this kind of market is shown in the photograph in Figure 2.4 of a local bar in Lanzarote. In this case the English tourist can feel safe, that upon entering 'paradise' they will find an English manager.

In the 'diversionary' mode, tourism acts as a diversion from the boredom and meaninglessness of everyday life, a state of alienation that is experienced in one's home society. Unlike recreational tourism, the individual's participation in tourism does not re-establish adherence to a meaningful centre in their home environment. Nor is this type of person looking to be centred in another culture, rather they are centre-less.

A tourist having an 'experiential' experience is involved in the process of searching for a centre away from their home environment. The behaviour of these tourists will reflect a much higher degree of contact with the local cultures than the previous two types, involving an exploration of their ideas and value systems. However, the experience-orientated tourist remains aware of the 'otherness' of the culture they are staying in, as Cohen (1979: 188) comments: 'the experience-oriented tourist, even if he observes the authentic life of others, remains aware of their 'otherness', which persists even after his visit; he is not 'converted' to their life, nor does he accept their authentic lifeways'.

The 'experimental' mode involves trying to live in different cultures with the aim of discovering a feeling of authenticity and a new centre. This type of experience is likely to have a higher level of involvement and immersion with the local culture than for the experience-oriented tourist. However, despite experimenting with different cultures, the tourist remains unconvinced by the authenticity of any particular one, and carries on searching. In the 'existential' mode the tourist has relocated

Figure 2.4 *'Paradise' is placed under English management*

their centre to another culture, but for practical reasons, such as work and family commitments, cannot physically relocate to their elected centre. They would therefore visit their elected environment as often and for as long as is possible within the constraints of everyday life.

Other typologies based upon tourists' behaviour have also been developed. In an earlier categorisation, Cohen (1972) recognised four main types of tourists, based upon the relationship between tourists, the destination environment, and tourism businesses. These are the 'organised mass tourist', the 'individual mass tourist', the 'explorer', and the 'drifter' as shown in Box 2.6.

Box 2.6

Cohen's (1972) typology of tourists

Category	Characteristics
The organised mass tourist	Highly organised travel; minimum contact with destination culture; travel in large groups
The individual mass tourist	Rely on the tour operator to arrange flights and accommodation; enjoy an element of liberty but will still tend to stay on the 'beaten track'
The explorer	Tries to avoid the tourist track; make their own travel arrangements; learn the language of the place they are going to and attempt to associate with local people; retain some of the values and routines of home life
The drifter	Attempts to become part of the local community by living and working with them; shuns contact with other tourists and the tourism industry

---

Cohen categorised the first two typologies, 'the organised mass tourist' and 'the individual mass tourist', as being 'institutionalised' tourists. The other two types, 'the explorer' and 'the drifter', he categorised as 'non-institutionalised' tourists. By institutionalised, Cohen implies a heavy dependence upon the tourism industry to organise their holiday. In contrast, non-institutionalised tourists place little reliance upon the tourism suppliers, searching for authentic experiences with an itinerary they choose.

Other researchers, notably Plog (1974), have tried to link personality characteristics to particular types of tourist. Plog's work is based upon the concept of psychographics, which has its origins in psychoanalysis. Its application to tourism involves examining and trying to comprehend the tourist's intrinsic desires to choose a particular type of holiday or destination. Plog's (1974) seminal work was developed in a commercial consultancy setting, working with airline business clients in the USA. Concerned with understanding the personality characteristics of people who were flyers as opposed to non-flyers, Plog used telephone-interviews to ask questions relating to personality characteristics. He subsequently established a continuum of typologies ranging from what he termed 'psychocentrics' to 'allocentrics'. The psychocentrics were found to display characteristics of being self-inhibited, nervous and non-adventuresome, whilst allocentrics showed traits of being variously experimental, adventurous and confident. According to Plog (1974), the population rests upon a continuum ranging from allocentrism to psychocentrism, along which it is normally distributed.

These personality characteristics were also found to influence the type of destination that would be sought by the tourist as shown in Box 2.7. Allocentrics are likely to enjoy the sense of discovery and new experiences of destination environments before other tourists arrive there. It should be stressed, however, that having an allocentric personality does not automatically imply a respect for the local environment. For instance, white-water rafting on an inaccessible river in the Himalayas would be a type of holiday matching the preferences of allocentrics, but does not necessarily imply a respect for the local cultural or physical environment.

Plog (1974) also linked the typologies to stages of destination development. As allocentrics tell their friends about their recent travels, these destinations become the latest fashionable or 'in' spots. The numbers of tourists increase, hotels and other tourist facilities are developed, and the destination orientates towards serving the mid-centric market. Tour operators are now likely to be bringing in groups of tourists to the destination. The allocentrics now begin to leave the destination as it has lost the characteristics of novelty and being unspoilt which attracted them there. At this mid-point in the cycle of development the destination is attracting the maximum number of tourists possible. If the destination continues to expand then it risks the problem of environmental degradation and ultimately a reliance on artificial or constructed attractions to attract tourists offering 'sun and fun'.

Box 2.7

The characteristics of psychocentrics and allocentrics

Psychocentrics

- Self-inhibited, nervous, non-adventuresome
- Search for familiarity and symbols of home, for example newspapers, drink and food, and an absence of foreign atmosphere
- Favour institutionalised tourism

- Prefer a high level of tourism development

Allocentrics

- Experimental, adventuresome, confident
- Prefer novel and different destinations and enjoy experiencing new environments

- Prefer non-institutionalised tourism allowing freedom and flexibility
- Prefer non-chain-type accommodation and few purpose-built tourist attractions

Source: After Plog (1974)

According to Plog (1974) such destinations are ultimately likely to appeal only to the small psychocentric market.

Besides the work of Cohen and Plog, numerous other typologies of tourists have also been created over the last twenty years. For instance, Dalen (1989), cited in Swarbrooke and Horner (1999), produced a four-group classification of Norwegian tourists. These included: the 'modern materialists' whose main motivation is hedonism and gaining a tan to impress people when they get home; 'modern idealists' who also like partying but are more intellectual than the previous group and want to avoid mass tourism and itineraries; 'traditional idealists' who demand culture, peace, heritage and security; and lastly the 'traditional materialists' who always look for special offers and low prices and have a strong concern with personal security.

Gallup (1989) also produced a typology of travellers for the American Express organisation. They identified the following: 'adventurers' who are motivated to seek new experiences from activities and other cultures and for whom travel plays a central role in their life; 'worriers' who display a considerable amount of anxiety about travelling and subsequently travel is not particularly important to them; 'dreamers' who are fascinated by travel but tend to orientate their trips to relaxation rather than anything too adventurous; 'economisers' for whom travel is not a central part of their life but travel because they need a break whilst seeking value for money; and finally 'indulgers' who like to be pampered and are willing to pay for a higher level of service when they travel.

Poon (1993) suggests that by the beginning of the 1990s a new type of tourism consumer had emerged. Changes in society, including consumer sophistication, heightened levels of environmental awareness, and increased familiarisation with travel, have led to the emergence of what Poon terms 'new tourists'. As Poon (1993: 9) comments: 'New tourists are fundamentally different from the old. They are more experienced, more 'green', more flexible, more independent, more quality conscious and 'harder to please' than ever before.'

Poon adds that the new tourists are spontaneous and unpredictable, and not as homogeneous as the old tourists. They are hybrid reflecting changing demographic patterns and lifestyles in society, as Poon (1993: 9) adds: 'Families, single-parent households, people whose children have left home, 'gourmet' babies, couples with double income and no kids (DINKS), young urban upwardly mobile professionals (YUPPIES) and modern introverted luxury keepers (MILKIES) are examples of lifestyle segments.'

Although typologies are useful in alerting us to the point that tourists are not homogeneous, there are a number of criticisms that have been made of the attempts to produce typologies in the field of tourism research. Sharpley (1994) criticises typologies as follows: for being static and not taking account of variations in tourist behaviour or experience over time; for being isolated from the realities of the wider social setting, for example, financial reasons may force a young family to take a package holiday although the parents may have allocentric tendencies; and for the lack of common agreement amongst tourism academics on the number of tourism typologies and how to clarify their boundaries which leads to repetition and a confusion of terminology.

Experiencing the environment

The various typologies of tourists suggest that tourists are not homogeneous and are likely to be searching for different kinds of experiences from the destination environment. Although there is a marked absence of research into how tourists experience the environment, Ittleson *et al.* (1976) drew up a taxonomy of environmental experiences that were later adapted by Iso-Ahola (1980) to the field of recreation. These modes of environmental experience have been expanded to the field of tourism to include a behavioural dimension as shown in Box 2.8.

This interaction demonstrates a gradation starting from little interest in the environment, beyond its providing the setting to indulge in a certain type of behaviour, to one of much greater interest and attachment. As with all taxonomies of experience the boundaries are not fixed and impermeable but are a reflection of the attitudes that a tourist may possess about the environment they are visiting. They are not exclusive and it is possible that an individual may experience more than one mode of interaction when visiting a destination.

It is likely that the attitudes of the tourist to the environment will be reflected in their behaviour. Tourists are likely to choose a destination environment because they feel it will be appropriate for the type of experience they are searching for. This means that the way environments are advertised carries implications for the behaviour of tourists and their subsequent impacts. For example, if environments such as mountain areas or coral reefs are advertised for activity-based kinds of tourism or as backdrops to high times and parties, then the possibility of a negative interaction between tourism and the environment is heightened. This scenario is exemplified in the following extract of an advertisement for a holiday in Tenerife from the 'Summer '99: Have it your way' brochure:

> We all piled into Bobby's the biggest club in Americas, where they had this brilliant English band and a rather (ahem) adult comedian – certainly broadened our knowledge if not our culture. The reps were all there too and really getting into it. Fuelled by that, we exchanged girls with hip strings for whales with blow holes on Atlantic Knights. What a mad day . . . a Freestyle cruise out to sea to spot dolphins and whales and other aquatic things. Oh and to drink a large amount of free beer.
>
> (Club Freestyle, 1999: 32)

Box 2.8

Modes of experience of the destination environment

Mode of experience	Interpretation	Behaviour and environmental attitudes
Environment as a 'setting for action'	The environment is primarily interpreted in a functional way as a place for hedonism, relaxation and recuperation. The physical environment may also possess the characteristics necessary for the pursuit of activities, for example, rivers for rafting, snow for down-hill skiing, coral for scuba-diving. The pursuit of satisfying the needs of relaxation or excitement and thrills is paramount over environmental appreciation. The destination environment is viewed as external to one's self	Conscious or subconscious disregard for the environment and a lack of interest in learning more about its natural or cultural history. In some cases a possible disinterest and disregard for environmental codes of behaviour leading to negative environmental consequences. Examples would include littering, breaking of coral, frightening of animals, disregard for local customs and traditions
Environment as a social system	The environment is seen primarily as a place to interact with friends and family	Physical setting becomes irrelevant as the focus of the experience centres on social relationships
Environment as emotional territory	Strong emotional feelings associated with or invoked by the environment which provide a sense of well-being. The surrounding environment is now an important part of the tourist experience in terms of personal development and is capable of producing deep-felt emotions	Sense of well-being and wonder at being in a different environment. The tourist may be moved to paint the environment, or write poetry about it, or sit or walk in contemplative wonder

Box 2.8 continued

Mode of experience	Interpretation	Behaviour and environmental attitudes
Environment as self	Merging of the physical and cultural environment with self. The environment ceases to be detachable from the person or external to them. The person's spiritual centre is now firmly located in this environment. Any damage or harm to the environment is perceived as damage to oneself. The destination environment may be experienced in an 'existential' way if physical constraints, for example, work or family commitments, restrict the opportunity for the person to relocate there permanently	Strong attachment to landscape(s) and culture(s) that are perceived as being 'better' than the home society. The tourist will have read extensively about their adopted cultural and physical environment and if necessary attempted to learn the language to allow them to interact with the local community in as meaningful a manner as possible

Source: Developed from Ittleson *et al*. (1976), Iso-Ahola (1980)

Although partying in the vicinity of whales and dolphins may not necessarily seem a particularly harmful activity, there is concern that too much human activity and noise stresses them. In New Zealand, Canada, South Africa and the United States swimming with whales has been banned. There are also initiatives being taken by organisations such as the International Fund for Animal Welfare to control whale watching, with limits on the noise and the number of boats, and restrictions on their proximity to animals (Harrison, 1998).

Similarly, if local cultures of destinations are displayed in tourist brochures in a way that relies upon stereotyping or shows a lack of respect, then the opportunities for cultural misunderstanding and conflict will be enhanced. For instance, in the copywriting of one tourism brochure in the early 1990s, Pattaya in Thailand was referred to thus:

> We'd guess there's getting on for a thousand bars in Pattaya . . . and then there's the nightclubs, discos, and restaurants. If you could imagine global TV announcing the end of the world in 24 hrs time, then what goes on in Pattaya all the time is what the approach to doomsday would probably be like. If you can suck it, use it, feel it, taste it, abuse it or see it, then it's available in this resort that truly never sleeps.

Given that tourists display different behaviour patterns, it is important for managing the environmental consequences of tourism that consideration is given in marketing strategies to attracting a type of tourist whose behaviour is likely to be compatible with the destination surroundings. This is an area of strategic marketing for tourism destinations which has to date been neglected with traditional market segmentation relying primarily upon geographical characteristics.

How the environment is used for tourism: an ethical question?

What is environmental ethics?

As discussed in the preceding section of the chapter, tourism is predominantly based upon the consumption of experiences through an interaction with environments composed of wildlife, nature and indigenous cultures. As Krippendorf (1987: 20) puts it: 'The countryside, the most beautiful landscapes and the most interesting cultures around the globe form the theatre of operations of this industry.'

Subsequently, tourism's interaction with these physical and cultural environments raises ethical questions of how they are used by the tourism industry and tourists. This ethical questioning of tourism is reflective of a broader ethical concern over our interaction with the environment, which has arisen from a growing awareness of environmental problems resulting from human actions. This awareness is demonstrated in a growing interest in general business ethics and associated issues, such as green consumerism, ethical investment and fair trade issues. Put alternatively, there is a wider questioning of the effects of consumerism on the environment in western society than ever before. In terms of what is meant by ethics, Goodpaster (1998: 51) defines ethics as follows: 'The study of ethics is the study of human action and its moral adequacy.' Tourism ethics could therefore be defined, 'as the study of the moral adequacy of the interaction between humans and the environment for the purposes of tourism'.

However, to date limited attention has been given to ethics in the field of tourism studies, as Lea (1993: 704) comments: 'It is rather surprising, given the importance of ethics in both the development studies and environmental literature, that so little space is devoted in the contemporary review of tourism studies to modern writing originating from this background.'

Nevertheless, the application of ethics to the environment has become an area of growing interest to philosophers since the 1970s. Vardy and Grosch (1999) recognise three categories of environmental ethics as described in Box 2.9.

Box 2.9

Three categories of environmental ethics

1 **The libertarian extension** The concept of individual rights is extended to all non-human animals and possibly inanimate entities. This means that all individual entities should be given the right to an uninterrupted freedom of existence. Some ecologists would extend the concept to include all ontological things (that is, beings and objects that actually exist). This interpretation means that a being does not have to possess an intrinsic value (alternatively expressed as consciousness) to have an ethical value.

2 **Ecological extension** Emphasis is not placed upon individual rights but upon the interrelatedness of all entities in the geophysical structure of the planet and their essential diversity. This approach is also referred to as 'eco-holism' and is exemplified in the Gaia hypothesis proposed by James Lovelock (1979).

3 **Conservation ethics** In this approach, emphasis is placed on ecological conservation for the benefit of humankind. The environment has an instrumental value in which it is not seen as an end in itself, but a means of gaining pleasure and profit. This is the most common form of moral reasoning over the environment and the most common dictate of environmental policy.

The concepts of 'conservation ethics' and 'libertarian extension' are also recognised by Simmons (1993), who deduces at least two possible meanings of the application of ethics to the environment. These are as follows:

1 An ethic for the 'use of the environment', which may be summarised as adopting an anthropocentric viewpoint over how we decide to use the resources of earth; and
2 An ethic 'of the environment', in which all non-human beings are given the same moral standing as that of the human species.

The first of these interpretations is probably the one most development decision makers would relate to, concerning the use of environmental resources for tourism. For example, a hypothetical ethical question could be, 'Is it ethically right for people to make their livelihoods through tourism, even though this involves the damaging and killing of coral reefs, thereby denying the next generation the right to use this resource to earn a livelihood?' The inference of this question expresses a human concern for the next generation rather than for the intrinsic value of the living organisms of the reef.

Simmons' second position is slightly more abstract, accepting that non-human beings possess a residual intrinsic goodness and moral equality with humans. Such a position would imply that the non-human world has 'equal' rights with those of the human world. The Romans, for instance, believed in *jus animalium*, the idea that animals possess rights independent of human civilisation. However, these rights did not extend to species of the wider non-human environment, with a great deal of deforestation of the Mediterranean basin taking place during Roman times. Also, directly associated with rights are responsibilities, and there is evidence to suggest that animals were tried in Courts of Law in the Middle Ages.

In contemporary times the dominant perspective on the environment is predominantly anthropocentric, concerned with enhancing benefits for humans, even if at the expense of the non-human world. Unfortunately for non-human life, humans are the dominant species, in terms of the hierarchy of nature. Given that in many situations humans deny rights to other humans, the chance of the human world acknowledging that the non-human world has intrinsic rights would seem to be minimal. However, at the end of the twentieth century there was a crusade by scientists and other intellectuals, to gain legal rights for great apes as described in Box 2.10.

Box 2.10

The 'Great Ape Project' – should they have 'rights'?

The rights of the great apes received increasing attention at the end of the twentieth century from a consortium of thirty-eight scientists and other intellectuals. The ultimate aim of the group was to obtain a 'United Nations Declaration on the Rights of Great Apes', in which apes would get: a right to life; a right not to suffer cruel or degrading treatment; and the right only to take part in benign experiments. The declaration would in effect stop zoos from keeping apes.

The basis of the campaign is that apes share 98.4 per cent of their DNA with humans and possess many of our characteristics, including self-awareness, the ability to form emotional bonds, basic linguistic abilities, and intelligence. Some scientists argue that apes also have the ability to recognise that someone else may have a different viewpoint. The campaign also suggests that the granting of rights could be extended to other non-human species, an example of the ethical position of liberterian extension. However the campaign has met with opposition from other quarters. A philosophical argument against giving apes 'rights' is that if they have rights, they are conferred with a status of moral beings like humans. This means that just as the person who kills an ape would be labelled a murderer, so an ape who killed another ape could quite legitimately be placed on trial for murder. The project has also been criticised on the basis that it sees nothing wrong with benign experiments to prove that apes are like us, and attempts to elicit elements of human and linguistic behaviour in animals who would not normally display it. Subsequently, one argument is that apes are still being treated as instruments of human purpose, evidence that the proponents of the project are not recognising the 'rights' of apes beyond those that humans are willing to give them.

Sources: Ezard (1999), Scruton (1999)

The demarcation of the human and non-human world and contemporary western attitudes to nature can be traced to the ancient Greek and Hebrew traditions, in which human beings are placed at the centre of the moral universe (Singer, 1993). Simmons (1993: 129) comments that there are two main traditions of the Christian religion: the first is that humankind is unique in being made in the 'image of God and therefore has the right to behave in a god-like manner to the rest of the cosmos'; and the second that 'humans are part of God's Creation just like the rocks and the trees and that no one part of this is inherently superior to another'.

Most commentators take the view that the dominant paradigm in western society is supported by the first tradition, in which dominion over the earth is paramount as a projection of empire mentality onto nature (Hooker, 1992). As Singer (1993) points out, God did drown almost every animal on earth apart from those that were led into Noah's Ark, to punish human beings for their wickedness. According to Nash (1989), Christianity is the most anthropocentric religion the world has seen, in which the earth is seen as a half-way house for trial and testing before death and hopeful admittance to heaven. Therefore in the Judaic-Christian religion, the earth does not embody god, instead removing god to on high. Singer (1993: 267) comments:

> According to the dominant Western tradition, the natural world exists
> for the benefit of human beings. God gave human beings dominion
> over the natural world, and God does not care how we treat it. Human
> beings are the only morally important members of this world. Nature
> itself is of no intrinsic value, and the destruction of plants and animals
> cannot be sinful, unless by this destruction we harm human beings.

In other cultures and religions the demarcation between humans and their surrounding environment is not as marked as in Christianity. For instance, as opposed to the 'frontier mentality' of the Anglo-Americans and the Anglo-Australians, the native Americans and Aborigines interpreted the natural environment not purely for its utilitarian value, but also as sacred space in which human interference required permission from the god or gods and often the performance of an appropriate ritual (Simmons, 1993). The presence of god was therefore recognised in nature, and as an integral part of the environment, a belief system known as 'pantheism'. It is therefore of little surprise that the behaviour of some tourists at the major tourist attraction of Ayers Rock in Australia has caused offence to the Aborigine people. Regarded as a sacred site by the Aborigines, tourists are encouraged to climb up it, whilst some even urinate upon it, an action akin to urinating openly in St Paul's Cathedral in London or in the Sistine Chapel at the Vatican in Rome.

Hinduism also has a long tradition of environmental protection under the concept of *ahisma*, meaning non-injury. The Hindu religion believes that harmony pervades the whole universe and has a respect for nature as a living being (Mieczkowski, 1995). In Buddhism, the taking of life is forbidden and vegetarian eating propagated, whilst Chinese Taoism stresses the harmony of the cosmos and the unity of all things, both human and non-human (Simmons, 1993). The Tao places emphasis on finding harmony with the cosmos by adapting to its ways and rhythms rather than trying to alter things and other people. In Islam, humans are stewards of the gifts of Allah, and this includes knowledge and understanding of environmental problems. However, even in other cultures such as the Indian in which the prevalent philosophy encourages a closer and integrated relationship between humanity and nature, the industrial development experienced under Prime Minister Nehru followed a western pattern of degrading the environment (Gosling, 1990). As Simmons (1993: 133) points out:

> It has to be said that in both East and West many religious traditions have collaborated with human behaviour that is destructive of species and habitat, and with non-sustainable development. In the West, obviously, there has been little sieving of technology and much talk of the conquest of nature; in the East no guidelines have been elaborated for alternative forms of economic and social growth that are ecologically sustainable. In all, some reconstruction of the historic faiths seems to be needed if they are to contribute to an evolutionary *modus vivendi*.

The dangers to the environment from less developed countries pursuing the goals of development encouraged by the West, are highlighted by Holdgate (1990: 91):

> However, the example set by developed countries to the developing world establishes individual wealth, consumerism, and personal prosperity as goals, although it is the pursuit of these values which threatens the same environmental degradation as has been caused in the developed world.

The separation of humans from the surrounding environment is also central to Kantian philosophy which tends to be heavily influential in western decision making (Ponting, 1991). The basis of Kantian philosophy is to treat nature as non-divine and to give humanity a spiritual freedom and domination over it. This raises three key points:

- the worth of nature is purely instrumental;
- no moral limits are imposed on humanity with regard to the use of nature;
- there are assumed to be no intrinsic internal limits within nature which we should respect.

Support to the dominance of man over nature has also been given by other philosophers, notably René Descartes, who believed that animals were both insensible and irrational and could not suffer. In his view, animals are mechanical like clocks, unlike humans who possess souls and minds. Such a philosophy encourages dualism by the separating of humans from nature and forms the basis for the use of nature to satisfy human desires and aspirations. Descartes encouraged the belief that humans were the masters of nature.

However, not all western philosophers were in agreement with Descartes; for example Benedict Spinoza believed that every being or object was a manifestation of a god-created substance. In his philosophy of 'animism' or 'organicism' there is a belief that a single and continuous force permeates all beings and things. Henry Moore, another important seventeenth-century philosopher, also believed that the spirit of god or *anima mundi* was present in every part of nature. In more contemporary times, the views of Henry Thoreau have been influential in questioning the way in which the environment is being used by humans for development. In the nineteenth century, Thoreau criticised the American obsession with 'things' and their treatment of nature in purely utilitarian terms. He was one of the first Americans to perceive the idea of inexhaustible resources as a myth (Nash, 1989). Thoreau recognised a god-like force in everything, and did not recognise any hierarchical organisation of nature, rather that there was a community of nature. He also associated abused nature with abused people. As Nash points out, such thoughts were unprecedented in mid-nineteenth-century America and Thoreau was virtually alone in holding them.

Another influential figure in aiding the conservation of nature in America was John Muir who championed the creation of national parks, leading to the establishment of Yosemite National Park in 1890, and two years later the formation of the Sierra Club to protect the park from development. However, as Nash points out, Muir did not go as far as Thoreau in recognising the rights of nature. His approach was based more upon a conservation ethic, interpreting nature as being valuable for people, in terms of recuperation, aesthetic appreciation and spiritual replenishment. In the early twentieth century, the questioning of nature being used in an instrumental fashion, and recognition of humans as part of the community of nature, was voiced by Aldo Leopold. Advocating the need for a 'land ethic', Leopold (1949: 219) comments: 'In short, a land ethic changes the role of Homo Sapiens from conqueror of the land-community to plain member and citizen of it. It implies respect for his fellow-members, and also respect for the community as such.'

Environmental ethics and tourism

As previously stated, the extension of the field of environmental ethics to tourism is an area that is currently underdeveloped, but is likely to become increasingly important as tourism continues to spread to encompass new environments. Given that tourism is dependent upon the use of physical and cultural resources, a central ethical question associated with its development is who will be the benefactors? As was discussed in Chapter 1, there are lots of different stakeholders who have different interests in tourism. This raises the potential for a clash of interests over tourism's development between the stakeholders, for instance between private developers and non-governmental organisations, or between governments and local communities. Such conflicts will not be equal, reflecting existing power structures and political hegemony, and sometimes involving the denial of democratic rights to certain groups. As Nash (1989) points out, certain groups of people benefit from the denial of ethics to either other groups or nature. For instance, in certain cases tourism may be developed to bring national economic benefits, such as employment and foreign exchange, against the interests and wishes of local people. Such a situation is described in Box 2.11 concerning tourism development in Malaysia.

This example of tourism development in Malaysia clearly raises ethical concerns over how tourism is used for development. The bringing of western tourists to indigenous cultures, such as the one described in Box 2.11, will undoubtedly bring with it cultural changes. These may be

---

Box 2.11

For the benefit of whom?

Commenting on the wishes of indigenous people in Malaysia not to be part of a tourism development plan proposed by central government, Malaysia's Deputy Minister of Tourism, K.C. Chan, comments: 'Do they (tribal peoples) want to sit in their longhouses (tribal dwellings) for ever or join a more advanced society? They are so used to their life in the jungle. If they can earn a better living from tourism, why not? It's part of the modernisation process, the 2020 vision. We should not be proud of backward people.'

Source: Vidal (1994)

particularly harmful to societies that have traditionally been based on mutual co-operation and dependence, and where competition over resources and the pursual of individual creation of wealth have previously not existed. The introduction of tourism into such societies can be divisive, causing jealousies and inequalities. That is not to say tourism should not be encouraged, as it can bring undoubted quality of life benefits to materially poor communites, including money to pay for health care, schooling and infrastructure development of piped clean water and electricity. The ethical question exists over the mechanics of the process of tourism development, and the extent to which there is any degree of local participation and democracy in decision making, as well as the extent of the economic benefits for the local community.

The development of tourism can also involve the denial of local people access to resources that they had previously enjoyed. For example, the development of safari tourism in Kenya, beach tourism in Goa in India and golf courses in Mexico and south-east Asia have all involved the exclusion of indigenous and local people from resources they have previously used for their livelihoods, as is discussed in detail in Chapter 3.

Another type of tourism, which raises ethical questions about the rights of women and children in particular, is sex tourism. The behaviour of western men in less developed countries, particularly the role of paedophiles who use children for sex, has raised protests from non-governmental organisations such as Tourism Concern. There is also concern that some destinations, such as Pattaya in Thailand, have become known as places to go to have sex with prostitutes. Such tourism raises ethical questions over the power relationships that exist within the indigenous society and also over the power relationships that exist between local people and foreign tourists. For instance, in Thailand many girls from the hill villages in the north of the country are sold by their fathers into the sex trade to help repay debts to money lenders. Perhaps even more ethically questionable, sometimes sex tourism may be developed as part of government policy. A newspaper report from the newsagency Reuters describes how the tourism authority in Cape Town is hoping to develop the sex industry to attract tourists. To quote: 'It plans a semi-regulated environment to ensure fair working conditions and a "first-class" service' (Reuters, 1999: 12).

Given that tourism affects fauna and flora, the ethical question of the use of the environment for tourism extends beyond the power relationships that exist between different groups in society, to include our relationship with the wider non-human environment of animals and other plant life. In the case of animals and their habitats, tourism has both the potential to protect habitats from other forms of development and human activities, such as poaching, and also the potential to destroy them. The potential destructivity of human action upon wildlife associated with tourism is graphically illustrated in Box 2.12.

Box 2.12

Tourists rush for 'kill a seal pup' holiday

The above title was the headline of an article that appeared in a British newspaper in 1993. The basis of the report was the response of thousands of Americans to a tour company who were advertising the opportunity to go and batter seal pups to death on the Newfoundland ice, under the marketing slogan 'Kill a Seal Pup' vacations. The executive director of the Canadian Sealers Association which was promoting the tours is quoted as saying: 'People want to come out and kill and it's a good market for us.'

Source: Evans (1993)

Such a direct reference to the culling of seal cubs would probably be regarded as unethical and a denial of animal rights by many in society. Perhaps this represents the consumption of the environment in its most direct form through tourism. In return for US $3000, tourists are buying the right to kill living animals, not in a dissimilar way to that in which tourism based upon game hunting continues to be encouraged in certain parts of the world. Although culling can be defended from an anthropocentric viewpoint by the need to sustain fish stocks, the selling of holidays based upon the aggressive instincts of tourists, raises ethical questions over human actions toward the animal world.

Concern over the ethics of tourism development has begun to induce a response from certain organisations within the tourism sector. In the early 1990s, the English Tourist Board (1991: 15) included as one of their principles of how tourism should be developed, the following statement: 'The environment has an intrinsic value which outweighs its value as a tourism asset. Its enjoyment by future generations and its long-term survival must not be prejudiced by short-term considerations.'

Such a statement adds a direct ethical dimension to tourism development, by recognising the 'intrinsic value' of nature, a type of value that is beyond the scope of economic evaluation but is recognised in its own right. The term 'intrinsic value' may seem abstract but can be interpreted as recognising that the non-human environment has a consciousness and therefore a value. A recognition of the intrinsic value of nature is likely to place emphasis upon the preservation of the environment and the exclusion of tourism. The other principles issued by the English Tourist Board, which relate to sustainable tourism, are discussed in Box 6.4.

Ethical concerns over the use of the environment for tourism have also begun to manifest themselves in policies for sustainable tourism developed by governments and the private sector, which are discussed in Chapters 5 and 6 of this book. In terms of a specific code of ethics for tourism, a significant development was the issuing of a 'Global code of ethics for tourism', by the World Tourism Organisation (WTO) in 1999. The code is established upon a range of policy instruments supported by the majority of world governments, aimed at protecting cultural and natural rights, some of which are generic and some which are specific to tourism. They include the following: Universal Declaration of Human Rights (1948); International Covenant on Civil and Political Rights (1966); Convention of the Rights of the Child (1990); Rio Declaration on the Environment and Development (1992); and the Manila Declaration on the Social Impact of Tourism (1997). The WTO ethical code contains a set of principles for how tourism should be developed, covering: mutual understanding and respect between peoples and societies; sustainable development; individual and collective fulfilment through tourism; tourism as a contributor to cultural enhancement; tourism as a beneficial activity for countries and communities; the obligations of stakeholders in tourism development; and the rights and liberty of tourism movement. Although it is not possible to reproduce the text of the code in its entirety, passages exemplifying the orientation of the text include the following.

> The understanding and promotion of the ethical values common to humanity, with an attitude of tolerance and respect for the diversity of religious, philosophical and moral beliefs, are both the foundation and consequence of responsible tourism; stakeholders in tourism development and tourists themselves should observe the social and cultural traditions and practices of all peoples including those of minorities and indigenous peoples and recognise their worth.

Tourism, the activity most frequently associated with rest and relaxation, sport and access to culture and nature, should be planned and practised as a privileged means of individual and collective fulfilment; when practised with a sufficiently open mind, it is an irreplaceable factor of self-education, mutual tolerance and for learning about the legitimate differences between peoples and cultures and their diversity.

All the stakeholders in tourism development should safeguard the natural environment with a view to achieving sound, continuous and sustainable economic growth geared to satisfying equitably the need and aspirations of present and future generations.

World Tourism Organisation (1999b)

At the beginning of the twenty-first century it would seem that ethical concerns over the interrelationship between humans and the environment are beginning to manifest themselves in the tourism sector. The extent to which these ethical concerns will influence the processes of development decision making is at the moment largely unknown and will only become evident with the passage of time. However, the development of ethical codes of conduct by internationally recognised organisations, such as the World Tourism Organisation, does suggest that ethics will have a more prominent role in tourism decision making in the twenty-first century than they had in the last century.

Summary

- The environment can be interpreted as the range of external conditions in which we live, incorporating physical, social, cultural, economic and political dimensions. The relationship between tourism and the environment therefore includes the influences in our own societies that encourage us to become tourists, as well as the interaction that exists between tourism and destination environments.

- Understanding why people become tourists is complex but is related to lifestyle issues of urbanised societies in advanced stages of capitalism. Tourism can viewed as a form of 'conspicuous consumption' relaying cultural messages about lifestyle and identity. Tourist behaviour in destinations will not be uniform, and will be representative of a mix of a range of individual motivations with wider cultural forces, which shape beliefs and attitudes.

- Tourism depends upon the existence of environments in other locations which are perceived as being attractive and desirable. This perception is socially constructed and subject to changes in fashion. Since the Industrial Revolution, the dominant perspective of what are attractive landscapes and cultures can be said to be more 'romantic' than 'classical'.

● The dependence of tourism upon the physical and cultural attributes of destinations raises ethical questions over who are the beneficiaries of tourism and the 'rights' of the non-human environment. This ethical questioning reflects concerns over the existing equality of relationships between different stakeholders in tourism and between the human and non-human environment.

Further reading

Bocock, R. (1993) *Consumption*, London: Routledge.

Krippendorf, J. (1987) *The Holiday Makers*, Oxford: Heinemann.

Nash, R.F. (1989) *The Rights of Nature: A History of Environmental Ethics*, Wisconsin: The University of Wisconsin Press.

Singer, P. (1993) *Practical Ethics*, 2nd edn, Cambridge: Cambridge University Press.

- The growing awareness of the effects of tourism upon the environment
- The negative consequences of tourism for the environment
- How tourism can benefit the environment

Introduction

As discussed in the last chapter, the rapid growth in demand for international tourism experienced in the second half of the twentieth century has resulted in the raising of ethical concerns over how the cultural and physical environments of destinations are used for tourism. This chapter considers the range of positive and negative consequences tourism can have upon the environment.

Changing perspectives on tourism's relationship with the environment

The reliance of tourism upon the natural and cultural resources of the environment means that its development induces change which can either be positive or negative. Today there is a growing interest in the environmental effects of tourism from governments, non-governmental organisations (NGOs), the private sector, academics and the public. This interest is reflective of a marked change in attitudes to our interaction with the environment that has occurred during the second half of the twentieth century. Society's concerns after the Second World War rested primarily with rebuilding the economies of Europe and environmental priorities were subsequently low. By the late 1960s, as the effects of the pursuit of economic growth upon the environment became more evident, environmental issues began to gain more prominence. The first break-up of a major oil-tanker, the Torrey Canyon, leading to the release of oil onto the south-west coast of England caused a high level of public concern and highlighted the fact that increased consumer consumption was not free of environmental risk. The increasing industrialisation of farming in the USA was also heavily criticised in Rachel Carson's (1962) book, *Silent Spring*, for the ecological damage caused by the use of agro-chemicals on farmland. This book subsequently had a major influence on public consciousness and eventually on regulatory policy, with the banning or restriction of use being placed on twelve of the pesticides and herbicides Carson identified as being most dangerous, including the notorious DDT. Academic interest in the environment was also heightened during the 1960s, and the disciplines of chemistry, physics, mathematics, ecology and biology began to be applied to the environment in an integrated manner within the field of environmental sciences.

However, tourism remained largely immune from environmental criticism, the image of tourism being predominantly one of an 'environmentally friendly' activity, the 'smokeless industry'. This perception was enhanced by the imagery of tourism, embracing virtues of beauty and virginity, as portrayed in landscapes of exotic beaches and mountain areas framed in sunshine. Nevertheless, there were one or two dissenting observations about the 'smokelessness' of tourism. Milne (1988) comments that in 1961 there was concern being expressed over the possible ecological imbalance that could result from tourism development in Tahiti in the Pacific. The observation of the effects of increasing numbers of people descending upon beautiful areas in the 1960s led Mishan (1969: 141) to write:

> Once serene and lovely towns such as Andorra and Biarritz are smothered with new hotels and the dust and roar of motorised traffic. The isles of Greece have become a sprinkling of lidos in the Aegean Sea. Delphi is ringed with shiny new hotels. In Italy the real estate man is responsible for the atrocities exemplified by the skyscraper approach to Rome seen across the Campagna, while the annual invasion of tourists has transformed once-famous resorts, Rapallo, Capri, Alassio and scores of others, before the last war no less enchanting, into so many vulgar Coney Islands.

By the 1970s people were becoming more aware and concerned over environmental issues. In 1972, the results of research by a group of scientists and business leaders into population growth, resource use and other environmental trends were published in the 'Limits to Growth: A

Report for the Club of Rome Project'. Predictions in the report of
pollution, resource depletion and heightened death rates resulting from a
lack of food and health services, raised public consciousness over
environmental issues. In 1979, the near meltdown of the nuclear reactor
at Three Mile Island in Pennsylvania, and the subsequent threat of a
major environmental catastrophe, alerted the public to the dangers of the
civil nuclear power programme. Opposition to nuclear power became a
central focus of the environmental movement in the 1970s, based upon
both the environmental consequences of the programme and its strong
link to the development of nuclear weapons. During the 1970s, questions
about the environmental impacts of tourism began to be raised more
widely, as tourism expanded internationally and the negative effects of its
development became more obvious. Recognition of the problems that
could be caused by tourism led the Organisation of Economic Co-
operation and Development (OECD) to establish in 1977 a group of
experts to examine the interaction between tourism and the environment.
Negative effects on the environment from tourism such as the loss of
natural landscape, pollution, and the destruction of flora and fauna were
already being noted. These concerns were also expressed in academic
circles, with the publication of *The Golden Hordes* by Turner and Ash
(1975), in which the whole process of tourism development was
questioned.

By the 1980s, global environmental problems resulting from the human
actions had begun to become popular media items. Global warming,
associated with the burning of the earth's carbon stocks for energy and
the associated release of CO_2 had become a major concern, as had the
depletion of the ozone layer. The predicted climatic changes associated
with global warming, and the increased risk of skin cancer resulting from
the depletion of the ozone layer, became matters of both interest and
concern to many members of the public. There was a growing
realisation that the pursuit of economic growth and increased material
consumption was having a profound effect upon our environment and
threatening the long-term well-being of humankind. Concern was also
being increasingly and vociferously voiced over the depletion of the
tropical rain forests of the world for agriculture and logging. Nuclear
power remained a concern as the world had its worst nuclear disaster at
Chernobyl in the Ukraine in 1986 and the effect of the nuclear fall out
was felt all over Europe. It is not a coincidence that in 1989 the Green
Party in the United Kingdom recorded 15 per cent of the vote in the
European elections.

During the 1980s, the spread of mass tourism beyond the Mediterranean basin into new areas, including south-east Asia, Africa and the Caribbean, meant that there was increasing focus on tourism as a form of economic development in less developed countries. Besides economic aspects of development, this focus also included concern over the environmental and cultural consequences of tourism development. The awareness that tourism could have negative effects was increasingly being recognised by NGOs. Pressure groups such as Tourism Concern, the UK-based campaigning group for humane tourism development, and the Ecotourism Society in the USA were established in the 1980s to promote ethically based tourism for both indigenous peoples and nature. Local pressure groups, concerned by the effects of tourism development on their physical and cultural environment were also formed, such as the Goa Foundation in India. The Goa Foundation have opposed tourism because of the loss of access to resources for local people and other associated human rights violations associated with its development, as is described later in this chapter. There was also evidence of increasing dissatisfaction by tourists with areas that were perceived as being overdeveloped or having lost their original attractiveness. For instance, Barke and France (1996: 302), commenting on the problems facing tourism in the Costa del Sol region of Spain in the late 1980s, state: 'Environmental decay and poor image have combined with overcrowding and low safety and hygiene standards, together with the popularity of cheaper forms of accommodation and catering, to reduce the perceived attractiveness of the region.'

During the 1990s, new environmental concerns became prominent, reflecting both local and global concerns. An ethical dimension was increasingly introduced into environmental campaigning over the rights of non-human life, with high profile and sometimes violent actions being taken for the liberation of animals from experimentation. Protests against road building became a central focus for environmental campaigners in Britain and other European countries as concerns over the loss of countryside and nature grew. The enthusiasm and popular support for this campaign led the British government to a major rethink of its strategy over transport, particularly the role of the private motor car and road building, leading to the cancellation of numerous road-building projects. Green politics in Europe gained increasing recognition through formal political routes in the 1990s, notably the formation of a governing red–green coalition in Germany, and by the end of the decade green politicians were in charge of the environment ministries of Germany,

France, Italy and Finland (Bowcott *et al.*, 1999). Concerns over the practices employed by farmers were also heightened with the outbreak of bovine spongiform encephalopathy (BSE) in Britain, which not only threatened animal life, but could also be transmitted to humans in the form of Creutzfeldt-Jakob Disease (CJD). Worries over genetically modified crops were also raised in Europe and there was a subsequent increased demand for organically produced vegetables, fruits and meats. Major protests were also made over the global inequality in world trade and the role of the World Trade Organization in encouraging the removal of trade barriers and import tariffs.

By the end of the 1990s, tourism development had for the first time been attacked directly by eco-warriors. Ski facilities were burnt down in Vail in Colorado at the beginning of 1999 because of their possible impacts upon wildlife. In the 1990s, the tourism industry began to take action over the environment, with many tour operators, hotels and airlines attempting to improve their environmental credibility. Growing concern over the use of the environment for tourism also found its way into the popular press; for example, the English newspaper the *Guardian* raises environmental issues associated with tourism almost on a regular basis in its weekend travel section. Tourism and the need for a more sustainable approach to its development also led non-governmental organisations not directly associated with tourism, such as the World Wide Fund for Nature, Voluntary Service Overseas and Oxfam, to become increasingly interested in tourism development. A growing number of tourists also became more interested to varying degrees in the environmental aspects of tourism. 'Green tourism', 'eco-tourism' and 'sustainable tourism', became favourite phrases in the industry. A summary of the changing attitudes of society towards the environment and tourism over the last five decades is shown in Box 3.1.

The environmental impacts of tourism

This section of the chapter discusses the environmental impacts of tourism. As will be illustrated tourism will have either negative or positive impacts upon the environment; rarely, if ever, will it have a neutral relationship with the environment. Although the discussion will attempt to be as holistic as possible, our knowledge of the impacts of much of human action upon the environment, and subsequently the amount we know about the effects of tourism, is limited. The limitations

of the discussion that need to be taken into account can be summarised as
follows.

- Research into impact studies is relatively immature and a true
  multidisciplinary approach to investigation has yet to be developed.
- Research into the environmental consequences of tourism tends to be
  reactive and therefore it is not always easy to establish a baseline
  against which to monitor changes.
- It is not always easy to separate out the environmental impacts
  attributable to tourism from the effects of other economic activities or
  anthropogenic factors, such as human habitation, and non-
  anthropogenic causes, such as natural environmental change.
- It is not always possible to separate the source of impacts upon the
  environment between local residents and tourists.
- The consequences of tourism are difficult to assess because tourism
  development is often incremental and the effects are cumulative.
- Spatial discontinuities are inherent to tourism. For example, the effects
  of air pollution caused by air and car emissions associated with
  tourism may contribute to acid rain which destroys forests hundreds of
  kilometres away.
  (Developed from: Hunter and Green (1995); and Mieczkowski (1995))

The impacts of tourism upon the environment can be separated into two
broad categories of negative and positive changes. To provide a structure
to the discussion, the first part of the following section deals with the
negative consequences of tourism and the second part with the positive
effects for the environment. This ordering is reflective of much of the
observation and commentary that has been expounded on the impacts of
tourism, which has raised awareness of the negative environmental
aspects that can result from tourism development, whilst the positive
environmental aspects are less well defined. This imbalance in the
literature is also a reflection of the fact that any change to the natural
environment by human action is likely to be viewed as harmful, involving
at the very minimum damage to individual flora and fauna, and
sometimes threatening the existence of species and whole ecosystems.
However, it should be realised that within the context of the discussion on
impacts, the extent to which we determine impacts to be either positive or
negative ultimately relies on value judgements. From an anthropocentric
viewpoint, these judgements need to be balanced with the consideration
that through the usage of natural resources for development, the standard
of living for humans can be improved. This is a particularly relevant
consideration for societies that live in material poverty and struggle to

Box 3.1

The relationship between society, the environment and tourism

Decade	Attitudes to the environment	Attitudes to tourism
1950s	Instrumental use for wealth creation to relieve the austerity of post-Second World War conditions	International tourism still restricted to a relatively small elite; high levels of participation in domestic tourism
1960s	Heightening environmental awareness; Torrey Canyon oil disaster; publishing of Rachel Carson's *Silent Spring* attacking 'environmentally unfriendly' farming practices in the USA; development of environmental sciences as an academic discipline	Beginnings of 'mass' participation in international tourism; development of the western Mediterranean; no environmental concern over tourism development
1970s	Growing awareness of pesticide and fertiliser pollution associated with farming; concerns over water pollution; publication of the Club of Rome report 'Limits to Growth' in 1972; awareness of global pollution and global warming in scientific circles; Three Mile Island nuclear power meltdown in Pennsylvania; formation of 'Greenpeace' in Canada in 1971	Increasing awareness in academic circles that tourism is not a 'smokeless industry' – mass tourism arrives in the eastern Mediterranean; Organization for Economic Cooperation and Development establish a working committee on tourism and the environment; publication of Turner and Ash's *Golden Hordes*
1980s	Major media and public concern over issues such as 'global warming', 'acid rain', 'ozone depletion'; Chernobyl nuclear power accident in the Ukraine; concern over the loss of the tropical rainforests; origins of green consumerism	Continued growth and spatial spread of tourism to south-east Asia and the Pacific; mass tourism in the Caribbean; by the end of the 1980s tourist arrivals began to fall to traditional locations such as the 'costas' of Spain, which were seen as *passé* and over-developed; tourism increasingly viewed as a development tool for less developed countries,

Box 3.1 continued

Decade	Attitudes to the environment	Attitudes to tourism
		founding of tourism pressure groups such as Tourism Concern (UK) and the Ecotourism Society (USA)
1990s	Protests against: development and road building; genetically modified crops; animal experiments; loss of rainforests; inequalities in world trade. On-going global concerns and an increased propensity to purchase organic food	'Eco-warriors' target tourism development in Colorado. More tourists becoming environmentally aware. The tourism industry begins to respond to concerns over the environment. 'Eco-tourism', 'green tourism', and 'sustainable tourism' become popular phrases

Source: After Hudman (1991)

meet their basic needs for food, clean water and shelter. Nevertheless, this is not to condone the negative effects of tourism upon the environment, as in many situations these result from a mixture of human ignorance and greed, rather than from a philanthropic desire to improve the living conditions of human beings.

The negative impacts

There are a broad range of negative physical and cultural environmental impacts resulting from tourism development, which can be categorised into three major types of concern: resource usage; behavioural considerations; and pollution. These are summarised in Box 3.2.

Resource issues and tourism

The development of tourism requires physical resources to facilitate its expansion. One of the most noticeable developmental aspects of tourism in generating and destination areas is airport construction. Airports are

Box 3.2

The negative environmental consequences of tourism

Issue	Problems	Examples
Resource usage: tourism competes with other forms of development and human activity for natural resources, especially land and water. The use of natural resources subsequently leads to the transformation of ecological habitats and loss of flora and fauna	Indigenous and local people can be denied access to natural resources upon which they base their existence and livelihoods. Land transformation for tourism development can directly destroy ecological habitats. The use of resources for tourism involves an 'opportunity cost' as they are denied to other sectors of economic development	• Airport construction in tourism generating and destination areas such as London and Malta uses large areas of farmland • Draining of coastal wetlands in Kenya for hotel developments • Loss of beach and coral reef ecosystems in the Caribbean • Deforestation of mountainsides associated with tourism in the European Alps and Himalayas • Lowering of the water table below the level of local wells as in Goa, India • Induced change to ecological habitats and a subsequent reduction in the number of species of flora and fauna as in Scotland and the European Alps • Exclusion of indigenous people from their land, such as the Maasai people from the Maasai Mari nature reserve in Kenya
Human behaviour towards the destination environment	Local people encouraged by the revenues to be gained from tourism and tourists may display ignorance and/or a disregard for the environment and indulge in inappropriate behaviour. This can lead to a range	• Disruption to eating and breeding patterns of wildlife animals in the Maasai, Kenya • Local people breaking off coral to sell to tourists off the Mombassa coast • Dynamiting of fish in the Amazon to

Box 3.2 continued

Issue	Problems	Examples
	of negative consequences for the physical and cultural environments	provide entertainment for tourists • Tourists walking over coral in the Caribbean • Increased crime, prostitution and drug taking in many destinations • Offence caused in Muslim cultures by western tourists wearing inappropriate dress to visit mosques and other cultural sites
Pollution • Water • Noise • Air • Aesthetic pollution	A range of different types of pollution can result from tourism. These can impact on different spatial scales from local to global. In destinations the effects of pollution are often associated with the level of tourism development and the degree of planning of implementation and environmental management controls	• Problems of human waste disposal generated by tourism in the Mediterranean and the Caribbean • Air pollution problems in the European Alps and the contribution of jet engine emissions to global warming and ozone problems • Noise pollution of air balloons in the Serengeti Park in Africa • Many coastal areas such as in parts of the Mediterranean and the Caribbean have had their coastlines transformed by standardised construction of tourist accommodation and are indistinguishable from each other

an essential part of the international tourism system and can generate major employment opportunities for local people. The expansion and development of airports is also beneficial to the tourist by offering easier access to a wider choice of destinations. However, the resultant effects for the environment, including the people who live near to airports, are not always as beneficial. For instance, airport development and expansion often involves the transformation of agricultural and recreational land which is covered by runways and terminal buildings. According to Friends of the Earth (1997), major international airports, like Heathrow in London, have paved areas equivalent to 320 kilometres of three-lane highways or motorways. In destination areas the extensive amount of land used for both airport and seaport development can also be problematic. For example, in 'small island developing states' (SIDS), the loss of agricultural land for airport and seaport development can lead to an increased reliance upon food imports to meet local needs (Briguglio and Briguglio, 1996). The development of an airport also requires additional infrastructure, such as new roads and railways, which again necessitates land-use changes and adds to the pollution of surrounding areas. As the demand for international tourism increases, the demand for the expansion of airports is likely to grow. According to Whitelegg (1999), air transport demonstrates the biggest growth rate of any form of transport and the International Air Transport Authority (IATA) estimate that the rate of growth in air passengers will continue at 5 per cent per annum until the year 2010.

Within destinations, the development of the tourism superstructure, such as hotels and attractions, and its associated infrastructure, also requires land. Tourism is often a competitor for land use with other economic activities, such as agriculture, and in some cases extractive industries like logging and mining. Subsequently, the use of land for tourism development is at the denial of other forms of economic activity, thereby incurring what economists refer to as 'opportunity costs', that is the potential economic benefits resulting from another type of development other than tourism are denied. The danger of unplanned and unregulated tourism development, in response to market conditions in which there is a high demand for tourism, can mean there is an overuse of resources for tourism and a lack of development of other forms of economic activity. This can lead to an economic overdependence upon tourism and a lack of diversification of the economic base. The danger of this situation is that if tourism demand to a destination decreases, there is a lack of development of other economic sectors to support the local economy. This is likely to lead to high levels of unemployment and associated social problems.

Another key natural resource that is essential for tourism is water. The addition of hundreds or thousands of bed spaces in a destination, combined with the lifestyle demands of western tourists, such as a daily requirement for showering, clean sheets and bath towels, means that tourism is responsible in some destinations for the consumption of copious amounts of water compared to the needs of the local population. Salem (1995) remarks that 15,000 cubic metres of water will supply 100 luxury hotel guests for 55 days, whilst the same amount will supply 100 nomads or 100 rural farmers for three years, and 100 urban families for two years. The effects of the development of tourism in areas where water resources are limited can mean that local people are denied the access to the water resources they previously used, for example to irrigate crops. They may find that streams previously used for irrigation have been diverted further upstream to service tourism development, or that the water-table level has been lowered by overextraction to service tourism establishments, rendering their wells useless. Where it is impossible to continue to extract enough fresh water locally, hotels can pay to have the water imported, whilst local people suffer water shortages. Unsurprisingly, access to water resources has sometimes led to conflict over tourism development between the developers and the local community, as exemplified in Box 3.3.

Box 3.3

Confrontation over water resources

In Tepotzlan in Mexico, the place from which Zapata led the peasant army in the Mexican Revolution of the early twentieth century, the residents who are mostly Nahua Indians protested against plans to build a golf course, five-star hotel and 800 tourist villas. Apart from the fact that such a development will be exclusive of local people in terms of employment opportunities, it is calculated that the development will use up to 525,000 gallons of water a day, threatening shortages in the town. Using slogans of 'Zapata lives' and 'Land and Liberty', the locals took over the town hall periodically, and barricaded the streets with barbed wire and boulders in protest against the scheme.

Source: Keefe (1995)

In Goa, in India, tourism development has also caused discord over resource issues between developers and local people. This has resulted in local people organising protest groups against tourism development and open antagonism towards tourists, as discussed in Box 3.4.

Box 3.4

Clashes over resources in Goa, India

The development of tourism in Goa raises many issues over the interaction between local people and tourism. Goa is a state in western India facing the Indian Ocean. It possesses the typical 'exotic' image of paradise for westerners with over 65 kilometres of sandy beaches and coconut palms. The area began to develop its international tourism (it was already very popular for domestic tourism) potential in the 1980s with the first German charter arriving in 1987. The development of tourism has been characterised by investment from outside the region and the building of large four- or five-star hotels. Unfortunately, this style of development has meant that tourism has excluded some local people from the resources that they need for their livelihoods. Nicholson-Lord (1993) gives examples of the 'Cidade de Goa Hotel' which built a 2.4-metre wall around a beach to deny local people access, and the Taj holiday village and Fort Aguada beach resort hotels, where guests are guaranteed water twenty-four hours a day whilst nearby villagers are denied access to the pipeline for even one to two hours a day. Many villagers face electricity and water shortages, with one five-star hotel consuming as much water as five villages, and one five-star tourist consuming twenty-eight times more electricity than a Goan. Many of the hotels were also built directly on the beach, damaging the dunes, and human sewage was put directly into the water without being treated. The extent of the feeling of exclusion of local people from the benefits of tourism has led to the growth of protest groups against tourism development, notably the Jagrut Goenkaranchi Fauz (Vigilant Goan's Army) and the 'Goa Foundation', an environmental group. Occasionally there has been open aggression towards tourists, such as the pelting of German tourist buses with rotten fish in the late 1980s, and ten tourists were beaten up by a group of villagers in Nuven village after knocking down two pedestrians.

Besides possibly having restricted access to water resources, local people may also find that they are excluded from other areas that they used to use for natural resources and recreation, such as beach areas. A typical sign on beaches that have been privatised to accompany up-market hotel development, especially in less developed countries, is like the one shown in the photograph in Figure 3.1, taken on the island of Langkawi in Malaysia.

Tourism development can also lead to the displacement of people from their homes, particularly of poorer people in less developed countries, many of whom possess no rights of land ownership and have limited or

Figure 3.1 *Privatisation of beach areas may result in local people being denied access to resources they previously used and enjoyed*

no access to legal representation. One of the most notable examples of the displacement of indigenous people associated with tourism was the exclusion of the Maasai people from their traditional lands, when the Maasai Mara reserve in Kenya was established as is detailed in Box 3.5. However, the exclusion and displacement of local people from lands for tourism development is not something that is specific to Kenya. The development of the Chitwan National Game Reserve in the Terai area of Nepal was also achieved by the exclusion of local people, and on the island of Langkawi in Malaysia, the compulsory reclamation of land by the state government led to the splitting up of a long established community and a loss of livelihood for many villagers (Bird, 1989). The case of Langkawi is discussed in detail within the context of environmental planning and management in Chapter 5, in Box 5.1. Displacement of people from the land for the development of 'golf villages' in South-East Asia has also occurred. The development of these golf villages usually involves the transformation of agricultural land, the development of condominiums, conference centres and other leisure facilities. Their size can be up to 80 times the size of a typical European golf course (Burns and Holden, 1995). It is difficult to have a global perspective on the numbers of people that may be displaced from the land for tourism development. However, it is evident that tourism development can be an exclusive process, and as landowners and entrepreneurs become aware of the financial opportunities to be gained from tourism, then the pressure on local tenants to leave their land is likely to increase.

Box 3.5

Coral reefs and game parks, issues of tourism development in Kenya

The country of Kenya in east Africa is endowed with natural resources that should make it an attractive tourism destination for decades into the future. It has both fantastic game reserves, a beautiful coastline with coral reefs, and cultural diversity. However, many of these resources are being put under threat from tourism development. The most popular area for holiday tourism in Kenya is along the coast, north and south of Mombassa. The coast has a diverse range of ecosystems including coral and mangrove swamps. However, the development of tourism has placed the coral under threat from sewage pollution from hotels, tourists walking on the coral and killing it, local people breaking it off to sell as souvenirs to tourists, and boats dragging their anchors through it. Mangrove swamps which form the basis of another vital ecological chain are also being cleared in an unsustainable fashion. They are being cut for construction poles, timber and fuel wood, as well as being removed for aquaculture farms. As a result of the demand by tourists for lobsters, crabs, prawns and fish, there has been overfishing and a decline in fish stocks, and subsequently lobsters and prawns now have to be imported from Tanzania.

Nor is tourism faring much better in some of the safari parks in Kenya, tourists being attracted by the 'Big Five' game animals: elephant, rhino, buffalo, lion and leopard. Indigenous people were displaced to create the parks in the 1940s, notably the Maasi people from the Maasi Mara park. This has radically altered their way of life and resulted in some of the Maasi having to migrate to the coast to sell their handicrafts to tourists. The consequences for the wildlife of increasing numbers of tourists in some of the parks has been the disruption of their breeding and eating patterns. Tourists are taken too close to the animals, elephants have been poisoned from eating zinc batteries thrown out in the rubbish from the tourist lodges that have been established, and there is also erosion of the vegetation from the excessive numbers of minibuses crossing the parks threatening the grazing of herbivores. Other problems in the parks include the pollution of water through lack of adequate sewage and other waste disposal facilities. This has led to wild animals and vultures feeding on the waste food and drinking from open sewers.

Sources: Visser and Njugana (1992), Monbiot (1995), Badger *et al.* (1996)

Besides adversely affecting the access of local people to resources, the physical environment can also be threatened by tourism development. As discussed in Chapter 2, a natural feature of many holiday destinations is the beach, but even it may be used as material for hotel construction ultimately leading to its disappearance. For example, in Antigua in the Caribbean, miles of pristine sand have been removed by sand-mining companies for use in constructing tourism projects, and sand has also been shipped to the Virgin Isles to build other beaches (Pattullo, 1996).

Similarly, coral reefs, the second most biologically diverse ecosystem on the earth after tropical rainforests and subsequently a great attraction for tourists, have come under threat from tourism development. Coral reefs are home to approximately 25 per cent of all marine species despite covering only 0.17 per cent of the ocean floor (Goudie and Viles, 1997). Their growth requires specific environmental conditions, including a water temperature of between 25 and 29 degrees centigrade, a relatively shallow platform of less than a 100 metres below sea-level to grow upon, highly oxygenated water, and areas free from sediment or pollution. There are three main types of coral reefs: 'fringing reefs', which connect directly with the shore; 'barrier reefs', which are separated from the shore by a lagoon; and 'atolls', which consist of a ring of coral reefs around a lagoon. The reefs are composed of a thin veneer of living corals that exist on the top of the skeletons of previous biological growths.

Coral is placed under threat from different aspects of tourism, including the construction of tourist facilities, inadequate sewage disposal measures to deal with human waste, and the behaviour of local people and tourists. Coral reefs have been mined for building materials in Sri Lanka, India, Maldives, east Africa, Tonga and Samoa (Mieczkowski, 1995). Besides being used for building materials, reefs may also be placed under threat from the dust created by construction which gets blown onto the reef, for example as in the Red Sea off the coast of Egypt.

The addition of untreated sewage into the water results in nutrient enrichment or eutrophication, which stimulates the growth of algae, which can cover coral reefs, in effect suffocating them. For instance, the discharge of partially treated sewage into the sea off the Hawaiian island of Oahu stimulated the growth of the alga *Dictyosphaeria cavernosa*, which overgrew and killed large sections of the reef (Edington and Edington, 1986). Similarly parts of the Great Barrier Reef off the coast of

Australia are being destroyed by the proliferation of 'crown-of-thorns' starfish, *Acanthaster planci*, which feed on the reef and have become abundant as a consequence of pollution and overfishing of their predators. According to Jenner and Smith (1992) millions of starfish have been observed, each one able to eat its own area of coral in a day, and as much as one-third of the Great Barrier Reef has been adversely affected to some degree. The Queensland tourism industry has been devastated, and the world's first floating hotel called the 'Barrier Reef Floating Hotel' operated for only a few months in 1988, before being towed back to Ho Chi Minh City from where it had come owing to a lack of tourist demand (Mieczkowski, 1995). The behaviour of local people and tourists towards the reefs can also damage the coral as is discussed in the next section of the chapter. One country in which tourism has resulted in negative effects upon its coral reefs, and other types of natural resources is Kenya, as is described in Box 3.5.

According to Pattullo (1996) over 90 per cent of the world's coral reefs are said to have been damaged by various forms of human activity. The extent of tourism's involvement in this damage is generally thought not to be large compared to other industrial sectors, although as Goudie and Viles (1997) point out, 73 per cent of the coral reefs off the coast of Egypt are thought to have been adversely affected by tourism. Nevertheless, the blame for the demise of the world's coral reefs cannot solely be laid at the door of tourism. Tourism is a contributory factor to the destruction of coral reefs along with other natural and human causes. These include storms and hurricanes, El Niño, overfishing, increased sedimentation produced by the deforestation of land and industrial pollution.

Hurricanes can kill living coral by flinging it up to the top of the reef, whilst El Niño which is an ocean current occurring off the coast of Peru every two to ten years, causes changes in ocean currents and sea temperatures. Changing sea temperatures have also been attributed to the effects of global warming. The raising of sea temperatures can threaten the food supply of coral, which is dependent upon microscopic algae called *Zooxanthellae* turning the sun's energy into sugars, to provide energy for the coral polyps. The algae are very temperature sensitive, and as the temperatures of the ocean rise the algae will leave the polyps denying them food, which may subsequently lead to the mass death of the corals. It is thought that with continued global warming and increasing ocean temperatures, this process of coral bleaching may occur more regularly. Additional threats for coral are also posed by more aggressive kinds of human actions. The reefs of French Polynesia in the tropical Pacific face the constant threat of nuclear bomb testing by the French, whilst the residual radiation remaining on some of the atolls from the American and British nuclear tests in the 1960s means they cannot be used for habitation or tourism (Mieczkowski, 1995). Industrial activity can also threaten the well-being of coral, for instance increased silting as a consequence of dredging for tin poses a threat to coral reefs in Thailand (Jenner and Smith, 1992).

The destruction of coral reef not only means a loss of biodiversity but also threatens other environmental resources upon which tourism depends. In the same way that tourism was likened to a spider's web in Chapter 1, major natural features such as coral, form part of interlinked and complex ecosystems. Coral not only provides an abundance of food for fish but also acts as a natural breakwater to help protect coastline and beach areas from erosion. Without this natural breakwater, beach erosion will occur at a much quicker rate than normal, especially where palm trees, the roots of which help to stabilise the beach, are removed to make space for hotel construction. The potential damage that unaware and unplanned tourism development can do to coral and beach environments is shown in Figures 3.2 and 3.3, developed from the work of Edington and Edington (1986).

Palm trees are perceived as being attractive and also help stabilise the beach

Coral reefs are the second most biodiverse ecosystem on the planet. They are a major tourist attraction and protect the coastline and beach from erosion by acting as a breakwater

The beach: a focus of contemporary tourism

Figure 3.2 *Pre-tourism development: the ideal paradise*

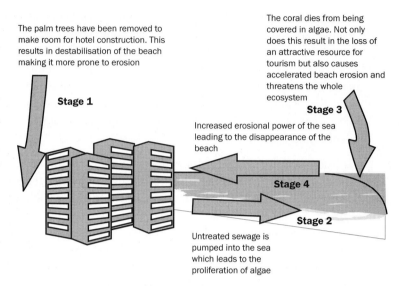

Figure 3.3 *Post-tourism development: the disaster*

An actual example of the type of construction mentioned in stage 1 of Figure 3.3 is shown taking place in Figure 3.4, which is a photograph of tourism development on the coastline of the Indian Ocean. The construction of tourist facilities would seem to be happening at a rapid pace, leading to the removal of palm trees, and destabilising the beach area.

Another important ecosystem in terms of making the earth a habitable place to live for humans, are coastal and inland wetlands, which cover approximately 6 per cent of the world's surface. Wetlands not only house a wide range of biodiversity and act as vast carbon-storing areas for the world, but also operate as a form of local flood control measure by absorbing vast amounts of water at times of high rainfall, and discharging it to adjacent areas in a slow and measured way. However, the restricted availability of land for tourism development in coastal areas, combined with the engineering ability to reclaim wetlands, means that wetlands are increasingly being used for tourism development. For instance, in the Languedoc-Rousillon region of France, five new coastal areas were developed on reclaimed wetland sites, in response to the lack of available land for building on the French Riviera (Klemm, 1992). One of the coastal areas, 'La Grande Motte', has subsequently become the third largest French resort in the Mediterranean after Cannes and Nice. The engineering project was enormous, involving the draining, dredging and

Figure 3.4 *Beach construction alongside the Indian Ocean*

filling of coastal marshes and saltwater lagoons, leading to the complete disappearance of the wetland fauna. A similar scheme to reclaim wetlands including mangrove swamps has taken place in the Caribbean, to accommodate the Sir Donald Sangster International Airport in Jamaica.

After coastal areas, the second most popular location for tourism is in mountain areas. Just as tourism has resulted in detrimental environmental effects in coastal areas, it has also caused negative impacts in many mountain areas. Mountain areas are physical environments which are sensitive to change, characterised by short growing seasons as a result of the cold temperatures, and soils which tend to be thin and nutrient deficient. The combination of these factors makes subsequent regeneration of damaged vegetation difficult.

Mountain areas have become popular destinations for activity-based tourism, and mountain systems worldwide are being used increasingly for activities such as downhill skiing and snowboarding, mountain biking, paragliding, white water rafting and trekking. One particular form of mountain tourism that is very popular, estimated to account for 3 to 4 per cent, or approximately 24 million of the recorded total of international tourist arrivals in 1997, is wintersports (World Tourism Organisation, 1998b). The economic benefit associated with sports such as downhill skiing and snowboarding has meant that their development has actively been encouraged through government policy in some countries, as a means of developing upland rural areas. Although the development of downhill skiing has aided the creation of economic prosperity in many mountain areas, it has also resulted in many negative consequences for the physical environment. The effects of downhill skiing in mountain environments are summarised in Box 3.6.

Box 3.6

The effects of downhill skiing upon the environment

Type of development	Processes	Results
Piste preparation	Removal of vegetation and boulders to a depth of 20 cm to allow snow accumulation	1 Ecosystem damage, for example loss of arctic-alpine vegetation 2 Visual Pollution – loss of aesthetic quality, especially in the summer
	Deforestation of the mountain sides	1 Increased avalanche risk 2 Increased propensity for mud slides 3 Disturbance of wildlife, for example Black Grouse in the French Alps, Ptarmigan and Red Grouse in the Scottish Highlands
Lift installation	Early ski areas built roads up the mountainsides to transport pylons	Ecosystem disruption – destruction of vegetation; disturbance to wildlife and loss of habitats
	Use of heavy cables to support tows and chairs	Death of birds colliding with cables, for example Black Grouse in the French Alps and the Red Grouse and Ptarmigan in the Scottish Highlands

Box 3.6 continued

Type of development	Processes	Results
Artificial snow-making equipment	Increasing use of artificial snow cannon, which involves great water usage, for example to produce one hectare of skiing surface requires 200,000 litres of water	1 Increase water usage – diversion of water and lowering of the water table 2 Energy consumption 3 Noise pollution 4 Use of additives to aid crystallisation of the water into snow leading to contamination of the soil
Increased infrastructure development	1 Building of extra roads to transport skiers 2 Hydroelectric schemes	1 Land-use transformation; noise and air pollution 2 Increased levels of salination causing loss of flora, for example the Australian Alps
Superstructure development necessary for destinations	Construction of hotel development and other usual amenities such as cafes, restaurants, bars	1 Land-use change 2 Air and water pollution

The development of mountain tourism requires the construction of hotels, apartments and associated infrastructure, placing increased pressure on land resources and animal habitats. The removal of trees to create ski runs, besides resulting in a loss of habitat for wildlife, also means that rainfall falling on the mountain slopes is not absorbed in the same quantity as before. The removal of trees causes a loss of cohesion and stabilisation of the soil by the tree roots, and subsequently the mountain slope is more prone to slippage. The combined effects of increased amounts of water running across the surface of the slope, its weakened stability and the force of gravity, has led to mountain areas becoming vulnerable to landslides. The effects of these landslides can be quite dramatic, sometimes involving the loss of human life. For example, Simons (1988) describes the avalanches of mud that swept down mountainsides during the summer of 1987 in northern Italy and southern Switzerland, in which 60 people died, 7,000 were made homeless and 50 towns and villages and holiday centres were wrecked. The cause of these landslides was attributed to the removal of mountainside forest for ski development.

Another mountain area to suffer deforestation and subsequent landslides as a consequence of tourism is Nepal. These landslides can be on a large scale, sometimes leading to whole mountainsides tumbling into valleys, as is shown in Figure 3.5. The landslide in this photograph occurred at Tatopani in the Annapurna area of Nepal, and was on such a scale it dammed the powerful Kali Gandaki River behind it, necessitating the dynamiting of the blockage to let the river flow.

A major cause of deforestation in the Annapurna area is the influx of mountain trekkers into the area during the last thirty years. The traditional resource used in the mountain valleys of the area for cooking and to heat water is wood. The continued growth in the number of trekkers to the area has put increased pressure upon forest resources as trekkers demand hot food and often a shower every day, as is explained in greater detail in Box 3.7.

Human behaviour towards the environment

An integral part of the tourism system is tourists and local people. The behaviour of both groups will be highly influential in determining the extent to which the consequences of tourism upon the non-human world are either negative or positive. The behaviour of tourists to the culture of the destination they are visiting will also be influential in determining whether tourism is viewed as a positive or negative force for change by local people.

A major natural attraction for tourists is wildlife but wildlife can be adversely affected by certain aspects of human behaviour. The viewing of wildlife species in their natural habitats has become an attractive activity for an increasing number of tourists, resulting in the intrusion of humans into environments which had previously been the exclusive preserve of wildlife. Ironically, the desire of the tourist to enhance their perceptions of nature by observing wildlife at close quarters can bring disruption to the natural behaviour of the wildlife they want to see. According to Duffus and Dearden (1990) cited in Roe *et al.* (1997), the extent of the impact of tourism on wildlife can be related to the type of tourist activity and the level of tourism development. Mathieson and Wall (1982) also

Figure 3.5 *A major landslide, resulting from deforestation associated with tourism, near Tatopani, Nepal*

Box 3.7

The environmental impacts of trekking in Nepal

Nepal is a small land-locked country situated between India and the
People's Republic of China and is one of the world's poorest nations with a
large foreign debt. It possesses outstanding natural resources for tourism,
notably the Himalayan mountain range, and is becoming an increasingly
attractive tourism destination for international travellers. The number of
international arrivals has risen from 6,179 in 1962 to 393,613 in 1996. One
type of tourism that Nepal has become renowned for is trekking, with 23
per cent of the tourists who go to Nepal participating in this activity. The
most popular trekking area is Annapurna, with 59 per cent of all trekkers
going there, compared to 19 per cent who go to the Everest Region and 9
per cent who go to Langtang. The Annapurna area is rich in biodiversity
containing both rare fauna and flora. However the arrival of over 40,000
trekkers per annum has put extra pressure upon the natural resources of the
area. Unplanned development has led to the construction of 700 tea houses
and lodges in which trekkers can stay. Large areas of the outstandingly
beautiful rhododendron forests that grow in the area have been removed to
provide timber for construction, and fuel to heat food for tourists, and
provide them with hot water for showers. Deforestation of the mountains
has resulted in tremendous landslides becoming a regular feature of the area
as shown in Figure 3.5. Waste disposal problems are also an issue.
Inadequate sanitation facilities results in pollution of water supplies, and
non-biodegradable litter is thrown into streams and piled in rubbish heaps
on the edge of settlements, again leading to pollution and hygiene
problems. Cultural problems have also been created in the area, with
children begging from trekkers, and the development of an inferiority
complex amongst some local people in comparison to the 'richer' tourist.

Source: Sparrowhawk and Holden (1999)

---

add that the resilience of wildlife to the presence of humans will
influence the degree to which tourism proves harmful to a particular
species. For example, the type of safari tourism practised in the Serengeti
Park on the Kenyan/Tanzanian border is representative of a highly
developed level of tourism, involving local operators taking tourists into
the park in minibuses, and animals being surrounded by thirty or forty
vehicles full of tourists taking photographs. The invasion of the territorial
space of the animals, and the associated increase in noise levels, raises
the stress levels of animals which is disruptive to their breeding and
eating patterns. For example, cheetahs and lions are reported to decrease

their hunting activity when surrounded by more than six vehicles (Shackley, 1996). The drivers of the minibuses are encouraged to ignore laws governing the proximity of their vehicles to the animals by the extra tips they receive from tourists for getting close to them.

Sometimes the threat to wildlife from tourism can be more direct, especially in communities where the level of environmental education is low, and an insensitive view of the environment is held by locals. For example, commenting on the backpacker operations that take tourists into the rainforest in Ecuador, Drumm (1995: 2) writes:

> Only 20% of local guides have completed secondary education, and very few are proficient in a language other than Spanish. Together with a social context which embues them with a settler frame of mind, antagonistic to the natural environment, the negative impacts of this operation are significant. Hunting wild species for food and bravado is commonplace during tours as well as the occasional dynamiting of rivers for fish. The capture and trade in wildlife species including especially monkeys and macaws is also very common.

Besides wildlife, other natural resources may be placed under threat from the action of local people. For instance, coral is damaged by local people who break it off to sell as souvenirs, as in the Bahamas and Granada, where rare black coral is made into earrings for sale to tourists. Local operators taking tourists out in boats to visit reefs sometimes drag their anchors through the coral causing localised damage, whilst tourists harm the coral by touching and standing on it. Additionally, shells are sometimes collected by local people to sell to tourists, as in areas of the Red Sea, the Caribbean and off the coast of Kenya. Key factors influencing the attitudes of local people to the surrounding environment, are poverty levels, economic opportunity and the extent of the provision by government and the private sector of environmental education.

Careless behaviour by tourists can also adversely affect wildlife and ecosystems. A common problem associated with tourists is littering, which can potentially result in the death of animals eating the litter, and also lead to the attraction of predators of endemic species into areas where they would not normally go. For example, elephants have been killed by eating zinc batteries thrown onto rubbish heaps surrounding the outskirts of lodges in the Maasai Mara in Kenya as described in Box 3.5. In the Cairngorm mountains in Scotland, foxes who are predators of the indigenous ptarmigan and grouse have been enticed into the area, leading to a decline in bird numbers (Holden, 1998). However, as Shackley

(1996) points out, although tourism can be detrimental to wildlife, the habitat of much wildlife is under a more widespread threat from other phenomena, such as agricultural development, urban expansion, and extractive industries like logging and mining.

The behaviour of tourists can also cause cultural changes in the societies they visit. One of the possible consequences for cultures that are exposed to tourism, especially in less developed countries where there is likely to be a marked difference between the lifestyle of the tourist and the local population, is what anthropologists term the 'demonstration' effect. According to Burns (1999: 101) the 'demonstration effect' applied within the context of tourism can be explained as:

> This effect refers to the process by which traditional societies, especially those who are particularly susceptible to outside influences such as youths, will 'voluntarily' seek to adopt certain behaviours (and accumulate material goods) on the basis that possession of them will lead to the achievement of the leisured, hedonistic lifestyle demonstrated by the tourists.

Seemingly there exists a desire for some groups living in non-western cultures, to imitate western lifestyles by acquiring the same symbols, a phenomenon also referred to as 'blue-jean' culture. The imitating of western lifestyles may however, go beyond the purely superficial, to appropriating the more materialistic and individualistic values of western society. The adoption of such values will undoubtedly cause change in societies whose previous cultural value systems are likely to have emphasised religion, co-operation, the family and the community. The copying of tourists' fashions can also lead to cultural conflicts between different groups in communities. For example, post-pubescent Muslim girls living on the Mombassa coast in Kenya who see western women dressed in bikinis, may wish to dress the same way, even though their religion dictates they should be completely covered. In societies with strong traditional values, such a situation is likely to result in conflict with the elders and men of the community.

Another criticism that is often made of tourism's effect upon local cultures is that it can lead to their commercialisation and commoditisation. This can result in established rituals and ceremonies becoming a parody of the authentic culture, to satisfy the demands of tourists, in effect turning into 'pseudo-events'. Tourists are unlikely to understand the significance and meaning of the event they are watching or participating in, and with the passage of time it is also possible that the

performers may lose sight of the original cultural importance of the
practice (Williams, 1998). Similarly, the same criticism can be made of
local handicrafts, which instead of being handcrafted for their cultural
significance, can end up being mass produced for sale to tourists. Yet as
Burns (1999) points out, it is often difficult to separate out cultural
changes resulting from tourism from those attributable to the wider
process of modernisation, such as the geographical spread of global
television networks.

However, the material wealth and lifestyle of the tourist can have the
effect of making one's own culture feel comparatively worthless and
limiting. What local people fail to understand, as no one usually informs
them, is that the behaviour of the tourist is atypical of how they behave
during the rest of the year. The fact that western society suffers from
increasing levels of family breakdowns leading to social isolation, high
crime rates, drug addiction and suicides and that there is a developing
underclass of impoverished and isolated people, are not points often
made to people in less developed countries. This was illustrated by
Marcel-Thekaekarka who had spent ten years working with the Adivasi,
the indigenous people of the Nilgiri mountains of Tamil Nadu in India,
on rural development schemes before she visited Germany with six of the
Adivasis. Although in material poverty themselves, the Adivasi were
shocked by the lifestyle of people in Germany, even though they were
materially much wealthier. The Adivasis expressed sympathy for their
lifestyles which seemed urbane and isolated as Marcel-Thekaekarka
(1999: 3) writes: ' "It's very nice to be here", Chathi, one of the six told
me. "But I couldn't live here. It's not my place. A man needs his family,
his community, his own people around him. Just money can't give you a
life. You'd shrivel up and die." '

Sometimes the problems that arise from tourism are what the majority of
societies would perceive as being socially unacceptable, involving crime,
prostitution and drugs. For example, the development of sex tourism in
Thailand and other parts of south-east Asia has resulted in increased
levels of Aids amongst the population. It has also led, as is the norm in
prostitution, to the exploitation of women by men.

Pollution

Pollution of the physical environment resulting from tourism can occur
on different spatial levels, including tourism generating and destination

areas, and other localities not directly connected with tourism but to which pollution is displaced. However, it is important to point out that tourism is a contributing factor to local and global pollution, along with a whole range of other services and industries. The pollution associated with tourism may be categorised into four main types: water, air, noise, and aesthetic pollution.

Water pollution

Water pollution is a major problem in many tourist regions of the world. For instance, in the most visited tourist area of the world, the Mediterranean, only 30 per cent of over 700 towns and cities on the coastline treat sewage before discharging it into the sea (Jenner and Smith, 1992). In the Caribbean Basin, where 100 million tourists annually join the 170 million inhabitants, only 10 per cent of the sewage is treated before being discharged into the sea. The most worrying aspect is that, compared to other areas of the world, these figures are actually good. Other regular international tourist destinations such as east Asia and Africa and the islands of the South Pacific, with a few exceptions, have either no sewage treatment or treatment plants that are totally inadequate for the size of the population (Jenner and Smith, 1992). The problem of water contamination from human sewage is not caused exclusively by tourism but is reflective of an inadequate infrastructure to meet the needs of both local people and tourists.

Besides the consequences it can have for human health, causing diseases ranging from mild stomach upsets to typhoid through the intake of water contaminated by faeces, human sewage also causes eutrophication (nutrient enrichment) of the water. This can pose a particular threat to coral and associated ecosystems as explained earlier in the chapter. Eutrophication of water may also lead to a downturn in tourism demand as experienced on the Romagna coast of Italy in 1989. According to Becheri (1991), the total number of tourist bookings on the Romagna coast fell by 25 per cent in 1989 compared to 1988, owing to the eutrophication of the Adriatic and the spread of algae on the surface of the water. The source of the pollution was from agricultural, urban and industrial wastes, which were deposited in the River Po and subsequently outflowed into the Adriatic. Similarly, the fear of a typhoid outbreak in Salou in Spain in 1988 resulting from contaminated water, led to a 70 per cent decline in tourist bookings the next year (Kirkby, 1996). By the late 1980s, many Spanish beaches were considered dirty, with only three beaches on the Costa del Sol being considered clean enough for Blue

Flag status in 1989 (Mieczkowski, 1995). The decline in water and beach quality contributed significantly to a slump in tourism receipts in Spain in the late 1980s.

Besides the pollution resulting from the inefficient disposal of human waste, water pollution is also caused by fertilisers and herbicides, which are widely used on golf courses and hotel gardens. The water containing the chemicals seeps through to the groundwater lying 5 to 50 metres below the earth's surface and through aquifers it eventually reaches rivers, lakes and seas (Mieczkowski, 1995). Other sources of water pollution are caused by motorised leisure activities such as power boating, and even suntan oil being washed off tourists when swimming can result in localised pollution. However, although tourism may seem to be a culprit of much of the planet's water pollution it is important to realise it is a contributory factor. The major sources of water pollution come from oil spills, industrial waste pumped into sea, and from chemicals used in agriculture.

Air pollution

A major source of air pollution within the context of tourism is associated with transport for tourism. Both air and car transport contribute to local and global atmospheric pollution through the burning of fossil fuels. The release of carbon dioxide (CO_2) is widely thought to be a major cause of global warming, and the emission of sulphur dioxide (SO_2) contributes to problems of acid rain which destroys forests and historic monuments such as the Parthenon in Athens.

Per passenger, aviation produces more pollution than any other form of transport, accounting for 3 per cent of the total amount of the world's carbon dioxide emissions or equivalent to the entire CO_2 output of British industry (Malone, 1998). The growth in air travel since the 1950s has been rapid, with the annual rate of passenger growth averaging 5 to 6 per cent per annum, over almost a 50-year period. Besides contributing to global warming, air transport emits 2 to 3 per cent of the total global emissions of nitrogen oxides, which are believed to reduce ozone concentrations in the stratosphere (Friends of the Earth, 1997). Emissions of nitrogen oxides and hydrocarbons at lower levels also contribute to regional smog problems by forming low-level ozone on calm summer days, which is harmful to health.

Air pollution is also associated with the development of airports for tourism. Health issues associated with airports include respiratory problems caused by emissions from aircraft and car traffic, and stress associated with noise pollution from air traffic. According to Whitelegg (1999) aircraft produce significant amounts of nitrogen oxides during take off and landing. He adds that Kennedy and La Guardia airports in New York are amongst the largest sources of pollution in the city, and that Midway Airport in Chicago generates more toxic pollutants than any other form of industry in the city. The effects of this pollution on health are dramatic, with aircraft engines being held responsible for 10½ per cent of the cancer cases in south-west Chicago caused by toxic air pollution (Whitelegg, 1999).

A common mistake is to equate transport in tourism solely with airlines, as it is often assumed that most tourism is undertaken by using air travel. However, this is not the case; for example, within Europe, the car accounts for 83 per cent of the total passenger kilometres (Cooper *et al.*, 1998). A very common pattern of summer holiday travel in Europe is for tourists from the countries of northern Europe such as Germany, Scandinavia, and the Benelux countries to drive down to the Mediterranean coast for their vacation. When domestic tourism is also taken into account, then the effect of the motor car becomes even more prominent, as the majority of domestic trips are undertaken by its use. For those people living in transport transit areas, the effects of tourism are predominantly ones of inconvenience, associated with pollution and safety concerns. Although the social and health effects of transit traffic upon local communities is an underresearched area, Zimmermann (1995: 36), commenting on transit traffic through the European Alps, remarks: 'The transit traffic is one of the most evident problems within the Alpine area. In several regions local populations' endurance levels have already been reached or exceeded.'

Within destination areas the air quality may deteriorate as a result of both extra traffic and construction. Dust generated during the construction of tourist facilities contributes to air pollution, for example Briguglio and Briguglio (1996) remark that the demolishing of existing buildings and the construction of new ones for tourism has generated vast amounts of dust in Malta. Yet, as with water pollution, tourism can also be adversely affected by displaced air pollution originating from sources elsewhere. Forests used for tourism in developed countries that are situated close to large industrial centres are particularly under threat from acid rain, caused by emissions of sulphur dioxide from coal-burning power stations. For instance, Mieczkowski (1995) refers to the Black Forest in Bavaria becoming the 'Yellow Forest', and Jenner and Smith (1989) comment that as a result of the damage to the forest caused by acid precipitation, there has been a loss of tourism income. It should also be noted that as tourism was a contributory factor with a range of other human activities to water pollution, the same is true for air pollution. The use of the motor car in everyday life for commuting and shopping, and the burning of fossil fuels to produce electricity, are larger air polluting activities than tourism.

Noise pollution

In psychological studies of humans, noise pollution has been found to affect behaviour in a detrimental fashion. According to Mieczkowski (1995) most complaints associated with tourism relating to noise are from air traffic. Noise pollution is particularly a problem for those residents who live around busy international and domestic airports, and the proposed development of airports may sometimes lead to violent opposition by local people and protest groups, as was the case with the construction of Narita Airport in Tokyo (Shaw, 1993).

Noise pollution from tourism will be particularly noticeable in destinations where tourists are searching for quietness and peace. Airflights in remote areas where quiet is expected, such as the Grand Canyon in the USA and the Himalayas, can cause disruption to tourists and recreationists. Noise pollution from the construction of tourism facilities can also be a problem for residents and tourists. Briguglio and Briguglio (1996) observe that intense noise is generated by the building of hotels and other construction activity in destinations. Night clubs open until the early morning, and increased car traffic from tourism movements, all add to the noise pollution experienced by both residents and tourists in tourism destinations.

Aesthetic pollution

The development of tourism facilities can also lead to a decline in the aesthetic quality of the environment. Commenting on the development of tourism in the Guadeloupe and Martinique islands situated in the Lesser Antilles, Burac (1996: 71) comments: 'The most worrying problem now prevalent in the islands relates to the anarchic urbanisation of the coasts. . . . Also, the built-up areas by the seaside are often not aesthetically attractive due to the diversity of architectural styles, the disappearance of traditional creole homes and the disorderly way in which public posters are displayed.'

Often tourism development is based upon maximising profits whilst ignoring aesthetic concerns. This has led to a uniform style of development along many coastlines of the world, that ignores local architectural styles, building traditions and materials as is shown in Figure 3.6.

In mountain areas tourism has also created unsightly development. Besides hotel and apartment construction, the development of ski lifts and pistes has also been heavily criticised as a form of aesthetic pollution. For instance, the Scottish Office (1996: 7) make the following remarks about the development of downhill ski facilities in Scotland: 'In addition [to adverse ecological effect], the infrastructure and uplift facilities associated with skiing can have a visual impact on what would otherwise be an unspoilt and undeveloped landscape.'

> **THINK POINT**
>
> Harm to fauna and flora resulting from tourism development is an emotive topic. Yet tourism can bring economic benefits to people who are in poverty, such as the money to buy medical supplies and food. To what extent is a reduction in biodiversity important (especially if aesthetic appreciation is not harmed) if economic benefits are being brought to an area from tourism?

Figure 3.6 *Tourism development has been criticised for producing a uniformity of structure that fails to reflect the building style of the local culture, as in this case in Lanzarote*

The positive effects

It is unlikely that any kind of human action has a beneficial effect for the natural environment it interacts with, other than to protect it from more damaging forms of human behaviour. Therefore, when we talk about the beneficial effects of tourism for the environment, we are in essence talking about tourism being used as a way of protecting the environment from possibly more damaging forms of development activity, like logging and mining. The exception to this situation is tourism's interaction with the built environment, where in post-industrial landscapes, tourism has been used as a catalyst to aid urban regeneration and improve the quality of the environment.

The development of tourism will, however, normally place an increased emphasis on the maintenance of a 'good-quality' environment in a destination, if tourism is intended to play a long-term role in the local economy. Yet the meaning of what is a 'good-quality' environment is highly value-ridden and debatable. Nevertheless, it is certain that the long-term economic success of tourism is often dependent upon maintaining a level of quality in the natural environment, which will satisfy the demands of tourists. As Mieczkowski (1995: 114) comments: 'The very existence of tourism is unthinkable without a healthy and pleasant environment, with well preserved landscapes and harmony between people and nature.' The consequences for tourism destinations that do not maintain a high-quality environment were illustrated by the examples of Salou in Spain, and on the Romagna coast in Italy, earlier in the chapter. This relationship between the economic success of tourism, the environment, and the tourist is shown in Figure 3.7.

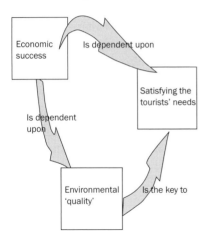

Figure 3.7 *The relationship between the natural environment, the local economy and tourism*

This diagram emphasises that the environment, including both its cultural and physical resources, is the key to satisfying the needs of the tourists and building long-term economic prosperity for tourism. It is therefore in the interest of the destination community to ensure that the landscape remains well preserved and that they provide stewardship of the environment.

Importantly, tourism can play a role in the conservation of the environment by giving it an 'economic value' through the revenues from tourist visitation. Given that development decisions are predominantly based upon an economic rationale, the revenues from tourism can help to protect habitats and wildlife from other more environmentally harmful forms of development, such as mining and logging, or from other forms of destructive human activity, such as poaching. The economic rationale of conservation through tourism is a theme explained in detail in the next chapter. Within the context of this section of the chapter, one of the most positive examples of where tourism has been used to conserve a particular species as described in Box 3.8 is the mountain gorilla project in Rwanda, which was proving very successful until the outbreak of civil war that took place there between 1990 and 1994.

One type of environment, in which some of the best examples are to be found of tourism making a positive contribution to environmental improvement, is in post-industrial urban environments. Although tourism can bring problems of traffic management, overcrowding and associated problems, such as littering to urban areas, it can also play in active part in the restoration of redundant industrial areas and of historic sights. Many of the post-industrial towns of the western world suffered the loss of traditional manufacturing industries, such as iron and steel, shipping and coal mining in the 1980s. Subsequently, tourism was turned to by many local governments as a catalyst for urban regeneration and as a means of providing new employment opportunities. The pioneer of using tourism to regenerate urban areas was Baltimore in the USA, where the decision was made in the 1980s to rejuvenate the waterfront through the development of shopping and recreation, to arrest inner city decline. Despite the numerous commendations that have been made of the redevelopment of Baltimore through tourism, Law (1993) criticises the success as being only skin deep, suggesting that poverty and housing problems still remain in the city.

However, there is little doubt that environmental improvements have resulted from the development of tourism in depressed urban areas. This is particular the case for the waterfront areas of major cities, such as Sydney in Australia, and Liverpool in England. One advantage of improving the environmental quality of the urban environment, especially where it is combined with improved infrastructure development, is that it enhances the image of the city and makes it more probable that other businesses and services will be attracted to relocate and invest there. Besides the advantage of attracting secondary investment, tourists who

Box 3.8

Gorillas in Rwanda

One of the most cited examples of how tourism can be used to aid conservation is in Rwanda in central Africa. The Parc National des Volcans in Rwanda is home to more than 300 of the world's estimated 650 mountain gorillas. They were particularly under threat from poachers, an export trade of gorillas' hands for ash-trays to the Middle East was one lucrative outlet, and also from the encroachment of agriculture leading to the removal of their habitat. The gorillas were popularised by the film *Gorillas in the Mist*, about the life of Dian Fossey who was an active researcher and conserver of the gorillas, ultimately resulting in her murder. There is little doubt that the film aided the development of tourism in the park. Tourist visitation of the gorillas is controlled by the Office Rwandaise du Tourisme et de Parcs Nationaux, and part of the revenues generated from tourism go to conservation agencies, notably the Mountain Gorilla Project and Dian Fossey Gorilla Fund. Visitors are taken out in small groups (maximum of eight) by well-informed local guides to see the gorillas in their natural habitat of dense bamboo forest. The total annual visitation to the park is between 5,000 to 8,000 tourists. Importantly, the project has been proven to be a financial success, critically benefiting local people, who in turn have taken a more active interest in conservation. However, the civil war in Rwanda between 1990 to 1994 unfortunately brought visitation to a standstill. In 1994 when 750,000 refugees were escaping from Rwanda to Zaire (now named the Democratic Republic of Congo), tens of thousands of people per day passed through the park, bringing with them their cattle and belongings. Many took refuge in the park. The occupation posed particular problems for the gorillas, including land-mines being placed in the forest, and gorillas being trapped in response to a lucrative black market for meat that had developed in response to famine. Between 1994 to 1998 rebel militias still held out in the park. The future of gorilla tourism in Rwanda therefore remains uncertain because of political unrest.

Sources: Mieczkowski, Z. (1995), Shackley, M. (1995), Shackley, M. (1996), Lanjouw, A. (1999)

come to the area also spend money, which induces a multiplier effect generating further demand for goods and services in the local economy.

The redevelopment of urban areas through tourism can also be aided by the development of tourist attractions which are rooted in the local heritage and history of the area. For instance, at Wigan in England, the town that was the source of inspiration for George Orwell's novel *The*

Road to Wigan Pier, the local municipality developed a heritage centre called 'Wigan Pier'. Reflecting the day-to-day life of an industrialised Wigan at the beginning of the twentieth century the centre attracts over 500,000 visitors per annum. Importantly the involvement of local people in establishing the centre helped regenerate local pride and interest in their heritage (Stevens, 1987). The development of heritage attractions to attract tourists to urban areas has now become a global phenomenon; for example Holden (1991) remarks that in Singapore a large historical and cultural theme park based upon the ancient Tang Dynasty of China has been developed, including 1,000 replicas of the Terracotta Warriors. However, the development of such attractions has led to criticisms of inauthenticity, especially where the theme seems to have little to do with the area it is situated in, or the attraction attempts to recreate life as it was in a historical period.

As with all forms of tourism, the benefits of urban tourism will only continue to be enjoyed if it is carefully planned and managed to ensure that areas do not become totally overcongested, as in the case of Venice, which is literally sinking under the weight of tourists. The influx of large numbers of tourists has been a contributory factor to a deterioration in the quality of life for residents in Venice, which has led to some of them leaving the city (Page, 1995). As for all forms of tourism, urban tourism has threshold limits beyond which the environment will be perceived as having declined in quality, with consequences for both local residents and tourists.

Summary

- The awareness of the environmental consequences of tourism has grown as society has become more environmentally conscious. Once perceived as being the 'smokeless industry' the expansion of tourism globally has led to an increased questioning of its environmental effects.

- Tourism can have negative impacts upon the environment. Major issues of concern rest over resource usage, pollution and aspects of tourist behaviour toward the environment they are visiting. The negative effects upon the environment include both physical and cultural aspects. However these negative effects must be offset against the economic benefits offered through tourism.

- Tourism can help protect the environment from potentially more damaging forms of development, such as logging and mining. It can have a particularly beneficial role in the regeneration of economically depressed urban environments.

Further reading

Hunter, C. and Green, H. (1995) *Tourism and the Environment: A sustainable relationship?*, London: Routledge.

Mathieson, A. and Wall, G. (1982) *Tourism: economic, physical and social impacts*, Harlow: Longman.

Mieczkowski, Z. (1995) *Environmental Issues of Tourism and Recreation*, Lanham, MD: University Press of America.

Tourism, the environment and economics

- The relationship between economics and the environment
- The relationship between tourism and economics
- How tourism can be used to conserve the environment using an economic rationale

Introduction

The fact that tourism can have negative impacts upon the environment, including the use of natural resources in an unsustainable fashion, the creation of pollution and the displacement of peoples, suggests that the wealth creation from tourism development is not universally beneficial. The processes of wealth creation and resource allocation are of particular interest to economists, and concern over the effects of development upon the environment has led environmentalists and some economists to criticise free market economics as being an inefficient mechanism of resource allocation. One means of wealth creation that is reliant upon the use of physical and cultural resources is tourism, and subsequently this chapter examines how the relationship between the environment and tourism is interpreted from an economic perspective.

The role of the physical environment

Most environmentalists and economists recognise that the physical environment provides certain key services for society as outlined in Box 4.1.

Within the context of tourism all of these services have a direct relevance. The physical and cultural attributes of destinations, provide the resources for wealth creation, by attracting tourists to them. Tourists

---

Box 4.1

Services provided by the physical environment for society

- Resources for wealth creation, based upon renewable and non-renewable types
- Ability to act as a waste disposal system assimilating the wastes of industrial production and other activities such as leisure and tourism
- Influences our well-being by providing us with aesthetic appreciation and landscapes for recreation
- Provides us with a life support system, that is, the combination of different ecosystems provide us with oxygen and water without which we could not exist

Sources: After Turner *et al.* (1994); Willis (1997)

---

desire to experience a 'good-quality' environment through participation in tourism as was discussed in Chapter 2, and there is little doubt that the experience of being in a good-quality environment endows tourists with a sense of well-being, through aesthetic appreciation and providing relaxing surroundings. The environment also performs the function of a waste disposal system for the tourism industry, for instance sewage is discharged from hotels into the sea, and aircraft release engine emissions into the atmosphere. At the same time the physical environment provides the communities of tourism destinations, and tourists, with the oxygen, water and food resources that are necessary for their survival.

It was the threat that economic development based upon free market principles was posing to these services, that led to criticisms of the ability of conventional economics to reflect the full environmental costs of production and consumption (Mishan, 1969; Turner *et al.*, 1994; Booth, 1998). Traditional models of economics are based upon continued growth, achieved by raising levels of production and consumption, on the assumption that this process will bring increased benefits to society. The process of consumption provides individuals with a sense of well-being or what economists refer to as 'utility'. However, this system of production and consumption fails to take into account the rate of depletion of environmental resources, and ignores the effects of pollution on resources that are not under private ownership, and therefore cannot be directly costed into the market system, such as the oceans and the atmosphere. Subsequently there is an inability to account for the long-term effects of consumption upon resource depletion, including the denial of the use of resources to future generations for their own wealth creation. This closed system approach, now challenged by many economists themselves, can therefore be criticised as failing to acknowledge the mutual dependency between the human and non-human environment.

A further criticism of the conventional economic model is that it ignores basic scientific principles, relating to the first and second laws of thermodynamics (Turner *et al.*, 1994; Booth, 1998). The first law of thermodynamics states that energy and matter can neither be created or destroyed, and the second, that when resources are processed the amount of useless energy (entropy) increases. A common form of entropy is pollution, and how to account for the costs of pollution upon the environment is a key question currently facing the discipline of economics. The major criticisms of conventional economics are highlighted in Box 4.2.

Box 4.2

Criticisms of the conventional economics approach to the environment

- Separates society's pattern of production and consumption from the physical environment and therefore ignores levels of resource usage
- Fails to account for the long-term effects of the depletion of natural resources
- Ignores the thermodynamic law of entropy and therefore fails to integrate the social costs of pollution into the market mechanism

---

Economic growth and human welfare

The phrase 'economic growth' has become a familiar one in contemporary society, and one that would seem to be politically desirable, although the effects of the processes used to achieve economic growth on the environment have led to increased questioning of how it should be achieved. The paradigm of 'economic growth' can be traced to the Cambridge economist Arthur Pigou, and the publication of his thesis, 'Economics of Welfare' in 1920. In his thesis feelings such as happiness and satisfaction were now referred to as 'welfare', and welfare could, according to Pigou, be expressed and measured in cash terms. Any satisfactions that could not be measured in cash terms via the market mechanism were to be ignored (Douthwaite, 1992). Pigou's analysis was therefore confined to 'economic welfare', alternatively expressed as happiness and satisfaction, that could be measured in monetary terms. One method of achieving happiness and satisfaction that is measurable in cash terms is through the consumption of goods and services. However, because the feelings of fulfilment and satisfaction we may gain from enjoying the environment do not have a measurable cash value, as a monetary exchange is not usually involved to acquire them, they have often been marginalised by decision makers.

Critically, Pigou also equated the amount of economic welfare with the national income per head thereby establishing a link between increasing levels of measured wealth and increasing levels of happiness and satisfaction, a paradigm that is still prevalent at the beginning of the twenty-first century. Consequently, one of society's primary concerns is increasing the 'standard of living' and maximising individual utility. The conventional view of the way to achieve this objective is through higher levels of production and consumption, commonly referred to as 'economic growth', on the premise that this will bring higher levels of welfare and happiness.

Given that many countries in the world are faced to varying degrees with a range of economic and social problems including high unemployment, lack of foreign exchange, poor health and education facilities, and growing populations, economic growth has subsequently been identified as a way of improving people's welfare. Many of the countries with the most extreme problems are usually categorised as being less developed countries (LDCs), based upon an economic comparison with the more developed countries of the West. The term 'West' is used here in a political rather than geographical context to indicate countries that meet the defined economic criteria of the United Nations to be considered as developed countries. This would include countries such as the USA, those of the European Community, and Japan.

The economic situation of many LDCs has been worsened by the dramatic fall in prices during the mid-1980s of traditional exports such as copper and tea, and the necessity to repay very high levels of foreign debt to the World Bank and other private banks. At the beginning of the twenty-first century it seems hopeful that the World Bank is willing to partially ease the debt burden of the LDCs. Much of this debt was associated with borrowing money for large-scale civil-engineering projects such as dam building for electricity provision, which were supposed to bring a promised panacea of 'economic growth' via the 'trickle-down' effect, that is, money being passed from the higher to the lower echelons of society. Large development schemes were also undoubtedly linked to political goals associated with national prestige. Loans were lent by governments and private banks to LDCs on the basis of the rising commodity prices that had been experienced until the mid-1980s. However, the effects of political corruption among powerful elites in many countries, and the failure of many schemes, mean that more than four decades after the United Nations First Development Decade of the 1960s, world poverty remains a major issue.

One form of economic development that has been encouraged by the United Nations to combat the problems of poverty since 1969 has been tourism. The use of Thailand as a 'pleasure ground' by the American Marines for recreational purposes, during the Vietnamese War, gave rise to the notion that tourism could be used in south-east Asia as a vehicle for economic growth. Tourism as a catalyst for economic development had already been successfully used by General Franco in Spain as discussed in Box 1.1. Tourism's ability to bring economic benefits and enhance political stability makes it an attractive development option to governments of countries whose environmental assets are conducive to the demands of western tourism markets, and where other options for economic development are few whilst large foreign debts need to be repaid. Following the example of Spain, a strong positive correlation has seemingly been established between the development of tourism and economic growth. Subsequently, from the perspective of government, the primary concern with tourism has been as a vehicle for economic growth.

The types of economic benefits that tourism can bring to destination areas are very well charted in tourism literature (e.g. Mathieson and Wall, 1982; Bull, 1991; Sinclair and Stabler, 1997; and Cooper et al., 1998), the most important ones from a national perspective being foreign exchange earnings, which are essential for the buying of necessities such as foodstuffs and medical supplies; reduction of the trade deficit; employment creation; increased expenditure and monetary flow; the strengthening of linkages to other sectors of the economy such as agriculture, fishing and construction; and to aid diversification of the economy from an overreliance on primary products. The strong link that exists between the development of tourism and economics is underlined by governmental responsibility for tourism often being given to the

Ministry of Economics or Trade, and tourism ministers being not infrequently graduates in economics. Owing to the fact that tourism can be developed with comparatively little financial investment in comparison to other kinds of industry, it is subsequently an attractive type of development for LDCs.

The emphasis of government policy for tourism has therefore traditionally rested with the economic benefits that tourism can bring, with relatively little consideration of the effects of tourism upon the environment. This is reflected in the economic analysis of tourism as Sinclair and Stabler (1997: 160) comment:

> The economics literature has concentrated on estimating income and employment generation and foreign currency earnings which developing countries gain from international tourism and the implications for their balance of payments. . . . The studies have not considered the impact of environmental deterioration on demand and thus income, employment and currency receipts, or estimated the social costs of tourism's role in economic development, either environmental or other.

The realisation that the increasing use of the environment for development can have adverse effects on human health and ecosystems has led to the questioning of the notion of economic growth, as a means of improving human welfare. Douthwaite (1992) points out that the terms 'standard of living' and 'quality of life' are frequently used interchangeably in contemporary society, whereas in fact they are different. The standard of living purely measures 'economic welfare', that is, satisfactions that can be measured in purely monetary terms, and therefore ignores a whole range of other factors that determine the quality of life. Such factors would include the quality of the environment people enjoy, levels of cultural activity, levels of health, and the chance to develop a rewarding religious or spiritual life. Doubts over the purpose of economic growth and the pursuit of an increase in the 'standard of living', which may bear little correlation to the 'quality of life', has led some economists to question how economic success can be best evaluated.

The degree of success in achieving economic growth is traditionally measured by key quantitative indicators, notably gross domestic product (GDP) and gross national product (GNP). Gross Domestic Product is the value of all the production of the goods and services in a country's economy. Gross National Product adds in the income received from residents of the country living abroad minus the money obtained in the internal economy which passes into the hands of persons abroad (World Guide 1997/8). Both GDP and GNP are taken by governments to be yardsticks of the success of economic policy, measures of the standard of living, and often seem to enjoy a somewhat revered status in the minds of ministers. An increase in the GNP in real terms over the figure for last year, ensures that governments can take some comfort and credit from their economic policy, whilst simultaneously reassuring the population that their 'standard of living' and therefore their well-being is increasing. A decrease in real GNP, commonly referred to as a recession, ensures that gloom develops from political commentators and that governments search for reasons lying within the global economy to explain the downturn in demand for goods and services.

Yet both GNP and GDP tell us little about the well-being of the environment that is being used to achieve economic growth, and subsequently have received substantial criticism from environmentalists, and some economists who have accused GNP of being a misleading indicator of welfare. For example, Pearce *et al.* (1989: 23) comment: 'But GNP is constructed in a way that tends to divorce it from one of its underlying purposes: to indicate, broadly at least, the standard of living of the population'; and Douthwaite (1992: 10) remarks: 'Since GNP only measures things which are bought and sold for cash, it ignores clean air, pure water, silence and natural beauty, self-respect and the value of relationships between people – all of which are central to the quality of life.'

A major shortcoming of the GNP measure is that pollution can contribute to its increase through expenditure on services to clean up pollution and associated health care costs. The costs of increased public spending to deal with crime in society also contribute to a growth in GNP. The term has subsequently been called *gross* national product by Porrit (1984). He adds that manufactured goods are deliberately not built to last, so that they have to be bought again, that is, they have a 'built-in obsolescence' that ignores the true costs of production to the environment.

Another criticism made of the GNP calculation is that the depreciation or cost associated with the loss of natural resources used for development is not included it. Subsequently, it is quite possible to increase the level of GNP whilst destroying the environment. For example, a country that is primarily dependent upon exporting hardwood for economic growth could be cutting down many trees and achieving a growth in GNP. However, the GNP figure fails to reveal the 'capital loss', that is, that the supply of hardwood may disappear in the next year or in five years' time because all the trees have been chopped down. Subsequently, attempts have been and are being made to change systems of national accounting to incorporate environmental costs and benefits. Such ideas include the 'System of integrated Environmental and Economic Accounting' (SEEA) and the 'Environmentally-adjusted net Domestic Product' (EDP), which attempt to include the costs of resource depletion and pollution in their methodologies. Nevertheless, as Bartelmus (1994: 35) comments: 'However, so far, there is no international consensus on how to incorporate comprehensively environmental costs and benefits in national accounts.'

Tourism and growth

The equation of growth with economic success also extends to tourism. For instance, the World Tourism Organisation rank countries in order, based upon the numbers of international tourists they receive and the total of their international tourism receipts, as is shown in Box 4.3. A ranking system usually infers that those at the top are the most successful, whilst those at the bottom are least successful.

Many ministers, particularly in less developed countries, view tourism as a means of pursuing economic growth and diversifying the economy away from an overdependence upon primary products, such as cash-crop agriculture and mineral production. The success of a tourism policy is usually evaluated by maintaining a growth in the number of tourists arrivals and associated expenditure, and obtaining an increasing percentage share of the total world tourism market.

In reality, the number of tourists and level of total expenditure can be misleading about the net economic benefits that tourism actually brings to a country or region. From an economic perspective, a much more accurate measure of tourism's worth to society is the amount of tourist expenditure retained within the local economy, the level of employment generation, and the equity of distribution of economic benefits. Besides the extra demand generated from direct expenditure by tourists in a destination, further income and employment is generated in the economy from the cyclical flow of money, an effect known as the multiplier effect. However, the amount of actual tourist expenditure that is retained in the

Box 4.3

The world's top tourism earners in 1998

Rank	Country	International tourism receipts (US $ million)	Market share % of world total
1	United States	74,240	16.70
2	Italy	30,247	6.80
3	France	29,700	6.70
4	Spain	29,585	6.70
5	United Kingdom	21,295	4.80

Source: World Tourism Organisation (1999)

Note: Success in tourism is often 'measured' by the growth in tourist receipts, visitor numbers, and international market share. However, such measurements fail to take into account factors such as economic leakages and the costs to the physical and cultural environments of the development of tourism.

destination economy will be determined by what is termed the 'economic leakage' from the total tourist expenditure.

The economic leakage of tourist expenditure begins before a tourist reaches a destination. In tourism, the method often used to purchase a holiday is via the intermediaries of travel agents and tour operators. However, not all the money paid by a tourist will reach the destination. Based upon the operations of one tour operator in the United Kingdom, the proportions of a tour operator's revenues that are typically received by destinations are displayed in Box 4.4.

From Box 4.4, it is evident that the majority proportion of the tourist's expenditure never arrives in the destination, what Smith and Jenner (1992) refer to as a 'pre-leakage' factor. A key determinant of the pre-leakage factor is the extent of the use of foreign tour operators and foreign airlines by tourists. For example, in a study of the package holiday market from the United Kingdom to Kenya, Sinclair (1991) found that for fourteen-night beach-only holidays, total foreign exchange leakages attributable to the overseas tour operator and foreign airline ranged from between 62 to 78 per cent. However, for safari holidays the

Box 4.4

Proportion of a tour operator's prices that is typically received by the destination

Region	Percentage
South America	45–50
Egypt	35–50
North India	35
China	30–35
South India	20

Source: Smith and Jenner (1992)

leakage factor declined to 35 to 45 per cent. This decrease is explained by the requirement of the Civil Aviation Board of Kenya that tourists taking internal flights to Nairobi, which is the usual departure point for safaris, must use state airlines. The extra expenditure on this service increases the percentage of the total tourist spend that remains in Kenya.

The tourism revenue that arrives in the destination will also be subjected to economic leakages. Put simply, the 'economic leakage' from tourism may be expressed as the quantity of revenue that fails to be retained in the economic system of the destination, from the total of the tourist expenditure. The factors that are likely to raise the level of economic leakage, and thereby reduce the economic benefits of tourism for local people, include: the degree of foreign ownership of the tourism industry and the subsequent remittance of profits to shareholders living outside the area; the servicing of foreign debts used to develop tourism facilities; importation of materials and manufactured products, such as fittings and furbishment for hotels, which can not be produced locally; and the importation of foreign foodstuffs to meet the requirements of tourists. The degree of economic leakage, will partly reflect the level of local economic development in a destination, and the extent to which goods and services required by the tourism industry can be supplied locally. For instance, if kitchen equipment has to be imported from Germany for a hotel in Turkey or butter needs to be imported from New Zealand for western tourists in Indonesia, the level of economic leakage will increase. Subsequently, the economic leakage factor for tourism is likely to be higher in less developed countries, who are likely to be more reliant upon imports and have a higher degree of foreign ownership of tourism facilities, than for developed countries. The World Bank estimates that on average, 55 per cent of gross tourism revenues received by countries, leak out (Smith and Jenner, 1992).

Not only does increasing the magnitude of tourism not necessarily equate with improved economic benefits but it can also result in increased environmental and social costs. Extra pressure is placed on the existing infrastructure, as well as the physical, cultural and built environments by increasing numbers of tourists. Increasing the numbers of international tourists arriving in a destination will for less developed countries most likely mean an increasing reliance on foreign tour operators and airlines to bring the tourists, and foreign capital to develop the required tourism facilities. The building of tourism facilities for profit also inevitably leads to uniformity of structure, and accusations of aesthetic pollution. Nor do the social problems associated with tourism, such as prostitution, drugs and crime, improve the quality of life for residents of destinations. Problems of crime will involve extra costs for government through increased expenditure on extra police officers and health care. The costs to tourism businesses and tourists therefore fail to reflect the full costs of tourism development, and are often externalised to third parties.

Problems of externalities

The development of tourism can lead to negative effects upon the physical and cultural environments, which can also be expressed as costs to society. For example, the pollution of the sea caused by sewage discharges associated with hotel development may lead to the contamination of fish stocks that subsequently stop fishermen earning their living, or extra air and car traffic at airports could lead to increased incidences of asthma in the surrounding population. Such costs may also be expressed in an inter-generational fashion, that is, the depletion of non-renewable resources, resulting from the pursual of economic benefits now, deny future generations the opportunities to meet their needs. With particular reference to tourism, Mishan (1969: 140) comments:

> For the cost [of the holiday] to the marginal tourist takes no account of the additional congestion cost he imposes on all others (tourists and inhabitants or of the additional loss of quiet and fresh air, or of the scenic destruction suffered by all in consequence of additional building required).

Such costs are therefore externalised by the decision maker and are subsequently termed 'externalities'. Willis (1997: 63) defines externalities as:

> These are consequences (benefits or costs) of actions (consumption, production of exchange) that are not borne by the decision maker, and hence do not influence his or her actions. Negative externalities that are by-products of people's use of the natural environment are commonly termed pollution.

According to Willis (1997), when externalities exist, market demand and supply curves no longer reflect all the benefits and costs of production. There exists, therefore, a discrepancy between the 'private costs' of production and consumption, and the full 'social costs' upon society. In this case the market mechanism fails to behave in an efficient fashion. Externalities arise because many environmental 'goods', such as clean air, water and landscapes, can be classified as 'public goods' sharing characteristics of collective consumption and non-exclusion. Collective consumption means that the consumption of the good by one person does not diminish the amount consumed by another person. Non-exclusion means that one person could not exclude another from consuming the resource, hence inferring there is an open access to it, and no price can be charged for it. Subsequently, for example, there is little economic incentive for airlines to contribute to the protection of the ozone layer and reduce their profit margins when it continues to be treated as a resource with zero cost. As Mieczkowski (1995: 160) comments in specific relation to tourism:

> They (meaning developers) treat the environment as an inexhaustible gift of nature because our free market system has been, so far, unable to incorporate the cost of natural resources and the value of environmental damage into the prices of tourism products. In other words environmental costs have not been internalised yet.

The realisation that external costs of production need to be internalised and reflected in the price of goods, with the aim of reducing pollution, led the Organisation for Economic Co-operation and Development (OECD) to adopt the 'polluter pays' principle in 1972 (Pearce, 1993a). The basic idea of the 'polluter pays' principle (PPP) is explained by Pearce (1993a: 41) as follows: 'The PPP requires that those emitting damage wastes to the environment should bear the costs of avoiding the damage or of containing the damage to within acceptable limits according to national environmental standards.'

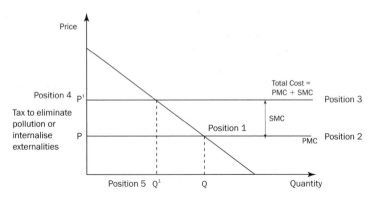

Figure 4.1 *The 'polluter pays' principle*

The implications of this approach for tourism businesses are that the costs of environmental damage would have to be internalised into their operating costs, which in turn would probably be passed on to the consumer, through augmented prices for the good or service as is shown in Figure 4.1.

To help explain Figure 4.1, an example is used of a hotel which pumps untreated sewage into the sea. In the existing market situation, position 1 represents the optimum level of usage of the sea by the hotel for sewage disposal as determined by free market forces. However, this situation results in pollution of the sea, causing other tourists and residents to suffer ill health, and lost economic opportunities for fisherman as a consequence of the contamination of fish stocks. In this situation a discrepancy exists between the private marginal cost (PMC) incurred by the hotel and their guests, shown in position 2, and the total costs of production shown in position 3. The difference between positions 2 and 3 is the social marginal cost (SMC), which in this scenario represents the costs borne by third parties from the effects of pollution, in this case the fishermen and other tourists and residents.

It is the gap that exists between positions 2 and 3 that the 'polluter pays' principle suggests should be closed. One way to rectify this situation is for the government to impose a tax on the hotel, or alternatively if a government has determined environmental standards for waste disposal that have been breached by the hotel, a punitive fine. This tax or fine could be estimated on the costs incurred for compensating fishermen for the lost commercial opportunities and for treating the illnesses of tourists and residents resulting from swimming in the sea. The imposition of a fine on the hotel means that the marginal external costs of the hotel's operations are now internalised to the hotel. The producer, in this case the hotel, is in effect now paying for the compensation of fisherman for their loss of opportunity to fish, and for the health care necessary to treat residents and other tourists. The hotel management may decide to pass these additional costs on to their hotel guests by raising their room rates, shown in Figure 4.1 as an increase from P to P^1, in position 4. In a price-sensitive and highly elastic market that typifies recreational tourism, the quantity of demand would shift from position Q to Q^1 (position 5), in effect meaning fewer tourists staying at the hotel and therefore less waste being pumped into the sea. This does not mean that untreated sewage does not continue to be pumped into the sea from the remaining guests, but that the level of pollution caused is now one that is socially desirable, that is, it involves a greater gain for the hotel and its guests than it does costs on the fishermen and other tourists or residents.

Similarly, Shaw (1993) uses the example of 'noise insulation' around airports to illustrate how the 'polluter pays' principle can be employed in connection with air transport. The costs of insulating houses against noise pollution from aircraft, for example with double glazing, may in some cases be recovered from airport authorities. These costs would be borne by the airlines and probably their customers, thereby making the polluter pay. However, as Shaw points out, such an arrangement does not compensate for the loss of enjoyment of gardens or open spaces caused by pollution nor for the annoyance factor when a window is opened in a house, although it can be argued that cheaper house prices around airports provide some compensation.

One of the problems faced with the 'polluter pays' principle is valuing the environmental resources that are being damaged and subsequently being able to estimate the cost of pollution. One way round this is for governments to establish national environmental standards, and the cost of meeting these standards should in the first instance be borne by the emitter of the waste. The polluter would then most likely pay through investment in waste management systems, or in some cases, for example if a hotel pumps an excessive amount of pollution into the sea as was the case in Figure 4.1, by a one-off punitive fine. Another method of operating the 'polluter pays' principle is through taxation, commonly referred to as 'green taxation'. According to Cairncross (1991), Pigou proposed the idea of taxation as a method of closing the gap between private and social costs, which he saw as the cause of environmental damage, back in the 1920s. However, as Cairncross points out, few countries have levied green taxes. One example of a green tax having been levied through tourism is at Cairns airport, Australia, where an environmental tax is levied on passengers using the airport.

Another problem of the 'polluter pays' principle lies in establishing the 'cause and effect' relationship. Shaw (1993) remarks that it may be difficult to always isolate the cause of an environmental problem, for example atmospheric pollution may be the result of emissions from many different sources. Distance factors are also a problem, as the source of the pollution may be hundreds or thousands of miles away from where the pollution manifests itself. A further problem highlighted by Shaw is that some adverse effects are difficult to reassess in an objective fashion.

Public or collective consumption goods

The fact that public goods are non-exclusive and therefore have zero cost means that a market for such goods cannot operate. Subsequently, they are at threat from overuse and degradation, which, expressed in economic terms, represents a social cost which may be borne by the owner of the resources or by society at large (Stabler, forthcoming). For example, this cost could fall on farmers or forest owners who are likely to achieve lower profits if they are forced to manage the land for its amenity value or on a community which has to pay increased taxes to mitigate the effects of river pollution.

In tourism, many of the negative impacts of its development are related to resources to which there is open access, and where there are no, or few, enforceable legal controls on usage. For example, the growth in the number of trekkers in the Annapurna area of the Himalayas has resulted in increased rates of deforestation of the mountainside. This in turn has led to increased landslides, and deforestation has also been associated with local climatic change. Sometimes the costs of environmental changes may be displaced to other areas. For example, as was described in Chapter 3, the deforestation of mountainsides for ski development in the Alps led to mudslides in northern Italy and Switzerland in which 60 people died, 7,000 were made homeless, and 50 towns and villages were destroyed.

The theme of overuse of public goods is discussed in Hardin's (1968) seminal essay 'Tragedy of the Commons', summarised in Box 4.5, which was important for encouraging debate about human interaction with the environment.

Box 4.5

'Tragedy of the Commons'

The likelihood of the overuse of common resources was highlighted in Hardin's (1968) seminal paper 'Tragedy of the Commons'. In his paper the example is used of the 'commons', an area of grass with common ownership, that is accessible to all the farmers in a particular community for grazing their cows upon. Assuming a steady state of equilibrium exists, that is, the use of the natural resource by the farmers provides them with an acceptable standard of living without there being any detriment to its quality, a sustainable situation exists. In such a situation the resource or 'commons' will be available for future generations to use to create wealth from.

However, in this scenario the addition of one extra cow will upset the balance. Therefore, if one farmer decides to increase his or her profit, by placing an extra cow on the commons, the balance between use and regeneration of the commons begins to be unsettled. Some farmers may decide that if they do not put an extra cow on the commons then another farmer (or 'competitor' in marketing terminology) will. The farmer may also reason that the cost of the use of the resources for the extra cow will be externalised, and spread amongst all the other farmers. In this situation there are now costs being experienced by other farmers, both short-term, for example maybe a slight reduction in the quality of the milk because the cows have difficulty getting access to enough grazing, and in the long-term because overgrazing could lead to the loss of the commons as a resource altogether. If all the farmers subsequently decide to adopt the same position, introducing one extra cow to the commons with the aim of maximising their profits whilst externalising their costs to other producers, the commons would become ultimately overgrazed. This would threaten not only the economic viability of the existing farmers but also the ability of the next generation to use the commons for milk production. In this situation, the true costs of production of the milk are reflected in neither the production costs nor consumer costs, because of a failure to incorporate the longer term costs of the loss of the resource. The market failure to reflect the total costs of production means that both the producer and consumer are benefiting in the short term, from not having to pay for the long-term environmental costs of production. The overuse of the commons will subsequently lead to the loss of the resource altogether to the detriment of society.

Source: After Hardin (1968)

The emphasis in Hardin's essay is placed on human selfishness, which leads to an overuse of the resources for the pursuit of private gain. An alternative way of expressing this concept is that the 'wants' of individuals go beyond their 'needs'. According to Max-Neef (1992) cited in Pepper (1996), fundamental needs are similar in all cultures and include subsistence, protection, affection, understanding, participation, creation, leisure, identity and freedom. Conversely, 'wants' are created by the advertising industry and are attached to the acquisition of material goods and services that provide enjoyment and a means of social differentiation, in other words consumerism. Green economists would argue that the natural resources of the earth can meet these 'needs' of the human population in a sustainable way, based upon ethical principles, but are unable to satisfy all the 'wants' of the human population.

The parallel between tourism development and the scenario portrayed by Hardin is a strong one. Probably no other type of development activity is as incremental as tourism in terms of its usage of resources for development. An additional hotel here, an additional flight there, all provide extra benefits for the suppliers and consumers of tourism through increasing profits and consumer choice. However, the development of facilities puts increased pressure upon the commons, typically ocean and mountain ecosystems, which may eventually lead to the destroying of the natural resources which attract tourists. The loss of the resources to attract tourists will ultimately lead to a loss of tourism and the inability of the local community to earn a livelihood from tourism as was suggested in Figure 3.7.

The solution of how to deal with the overuse of the earth's commons (e.g. oceans, air, natural parks) is not clear. The issue is complicated by there being different types of natural resources based upon characteristics of ownership, exclusivity, property rights, and agency and management structures (Stabler, forthcoming). For example, national parks have a much clearer agency and management structure than the global atmosphere, and in spatial terms have defined geographical boundaries. Subsequently, controlling access to the park is much easier than controlling access to the atmosphere.

Some critics of Hardin's (1968) work question his assumption of a finite carrying capacity. They place their faith in new technology and environmental design to extend its capacity. Others dispute Hardin's view of public attitudes to the use of the commons, arguing that common users were never oblivious to the common good (Pepper, 1996). For example in England, herdsmen used to consult each other over the possible expansion of herds to ensure that no threat was posed to the sustainability of the commons, because it was in their interest to do so.

Tourism, economics and conservation

Criticism of the failure of conventional economics to recognise the impacts of growth upon the environment has led some economists to examine how economics can be more inclusive of the environment, and subsequent attempts have been made to develop methodologies to give a value to the environment. One method is based upon expressed consumer preference for the environment, although to many environmentalists the idea of determining the environment's value in such a way may seem abhorrent, failing to recognise its intrinsic value. As Cairncross (1991: 43) comments: 'Nothing so annoys environmentalists about economists as their attempt to put a price on nature's bounty. "What am I bid for one ozone layer in poor condition?" they tease. "How much is the spotted owl worth?"'

Putting an economic value on environmental assets is difficult but, as Cairncross continues to point out, society puts values on environmental assets all the time by deciding which policies to pursue. The reality is that ethical arguments about the 'rights' of nature seem to have little credence in development decision making, government policy advisers are more likely to be trained economists than environmentalists or philosophers. It is therefore important to demonstrate the economic value of environmental assets in their existing form as accurately as possible. As Cairncross (1991: 43) writes, 'In a world where money talks, the environment needs value to give it a voice'; and Pearce (1993a: 15) comments, 'If, on the other hand, conservation and the sustainable use of resources can be shown to be of economic value, then the dialogue of developer and conservationist may be viewed differently, not as one of necessary opposites, but of potential complements.'

The willingness of individuals to give money to, and become members of, non-governmental organisations aiming to protect the environment, such as the World Wide Fund for Nature (WWF) or the World Society for the Protection of Animals (WSPA), suggests that the worth of the environment can be expressed in economic terms (Pearce, 1993a). According to Turner *et al.* (1994), one method of aggregating individual preferences, or measuring the gains and losses of well-being is by looking at what people are 'willing to pay' for something, the most common measuring rod of preferences being money. Turner *et al.* (1994: 94) comment: 'A measure of an individual's preference for a good in the market-place is revealed by their willingness to pay (WTP) for that good.'

The level of willingness to pay can be assessed through direct questioning: 'What would you be willing to pay to ensure the survival of a lion as a species?'; or 'What would you be willing to pay to protect the ozone layer?' This technique, based upon asking the beneficiaries what they would be willing to pay for the environmental benefit under consideration, is called the contingent valuation method. As Pearce *et al.* (1989) point out, the temptation would be to respond that such species or environmental assets are 'priceless'; however, it is unlikely that many of us would be willing to give all our worldly possessions to preserve the lion. So in this sense there is an economic value attached to its preservation. One study that tried to estimate the economic value of the lion was conducted in the Amboseli National Park in Kenya, which found that each lion was worth $27,000 per annum in 1980s values, expressed in terms of visitor pulling power. The study also demonstrated that wildlife tourism was economically preferential to the other main development option of agriculture. The parks net earnings from tourism were found to be $40 per hectare per year, fifty times higher than the most optimistic projection for agricultural use (Boo, 1990).

Another method of calculating the value of the environment is an indirect method, the 'travel-cost method', which as the name suggests is calculated on what people spend on travelling to recreational sites such as national parks. It is indirect because it uses surrogate pricing, in this case travel costs, as a measure of value. The travel-cost method (TCM) was developed by Clawson in the United States in 1959 and perhaps represents the most widely accepted and used method for the valuation of recreational sites (Coker and Richards, 1992). The underlying presumption of the TCM is that the costs incurred in visiting for example a national park reflect to an extent the recreational value of the park to an individual. Information on the expenditure to a site is usually gathered as part of a visitor survey. The TCM is essentially measuring the expectation of enjoyment of visitation before the visit takes place, as the visitor or tourist has to commit themselves to the cost of the trip to get there. In the case of tourism, given that the desires of tourists discussed in Chapter 2 related closely to vacationing in a high quality environment, it would seem that tourists place a 'value' on the quality of the destination environment by spending money and time to travel there.

Although the revenues that can be obtained through tourism can help protect the environment from other more environmentally destructive development alternatives, an important caveat of this line of argument is that if natural habitats or wildlife are judged not to have sufficient economic value in comparison to other development options, then a pretext is set for their removal. Shackley (1996) also highlights another danger of the economic valuation process with particular regard to wildlife. There are many species of wildlife which are not attractive to tourists in the dramatic sense that elephants and lions may be but have a role to play in the ecological system of the area. By their lack of inclusion, or the lack of ability to be able to value them because they are not on the tourists' itinerary of animals to view, their value will be undermined in economic terms, placing their continued survival under threat. Shackley (1996: 127) recommends the following approaches to valuing wildlife:

- calculating total gate or licence fees to estimate the value of tourism in a particular destination;
- estimating visitor expenditure on equipment, lodging, food and transport within a designated area;
- looking at employment generated.

Certainly in areas where tourism takes place in spatially defined areas, such as national parks, that possess an evident management structure, then the use of such methodology will be easier to implement than in areas which are less well defined. Where tourists wish to see 'wildlife', then it is possible to give the wildlife a market value, based upon the willingness of individuals to pay to see it. This is exemplified in the comparative study of three national parks in different countries shown in Box 4.6.

THINK POINT

How can tourism be used to aid environmental conservation? What are the dangers of giving wildlife a market value as a means of arguing the case for their conservation?

Besides methodology aimed at calculating the value of the environment, policy instruments can also be adopted to help give recognition to the value of conserving environments. One way of aiding the conservation of natural resources in less developed countries is through 'debt for nature

Box 4.6

Calculating the 'willingness to pay' of visitors to national parks

The Department for International Development (DFID), in the United Kingdom, commissioned a study to examine the amount of possible revenue that was being lost from three national parks in three different countries (the Komodo National Park, Indonesia; the Keoladeo National Park, India; and the Gonarezhou Park, Zimbabwe). In all three parks the price of entrance fees to the parks were not established by the market and it was suspected that visitors would be willing to pay considerably more to visit them. The extent of this user surplus was assessed using the 'contingent valuation' method, which explored the response of visitors to hypothetical rises in entrance fees. For each park, questions concerning tourists' 'willingness to pay' were included in tourist surveys. The following results were found:

Proposed entrance fee	Proportion of sample willing to pay (%)		
	Gonarezhou	Keoladeo	Komodo
Current	100	100	100
X 2	79	91	93
X 4	24	70	81
X 8	8	*Absent*	62

Interviewees were surveyed in the park and asked how much they would be prepared to pay for the current experience. The results revealed that price elasticity (defined as the ratio of fractional or percentage change in demand to the fractional or percentage change in price) is appreciably more elastic at Gonarezhou than Keoladeo or Komodo. However, as the authors of the report point out the entrance fee was already five to seven times higher at Gonarezhou than at Keoladeo or Komodo. Critically the authors also found that there is a significant difference in the willingness to pay amongst different types of tourists. For example, in the analysis of Komodo National Park members of conservation groups and older people were likely to be willing to pay higher entrance fees than any other kinds of tourists. However, the authors continue to point out the restrictions of the willingness to pay method, that is, people are being asked how their behaviour would change in a hypothetical situation, and there is no guarantee that in reality they would behave like that. However, the case study indicates that there is a potential to raise income through raising fees, and that any tourism strategy aimed at maximising revenue needs to take into account the types of visitors it wishes to attract, not interpreting the tourist market as being homogeneous.

Source: Department for International Development (1997)

Box 4.7

Debt-for-nature swaps: the case of the Central Region of Ghana

The Central Region of Ghana possesses a rich cultural and wildlife base. Besides its indigenous culture, it was the first area of contact between European governments and the people of West Africa. Consequently, the remains of castles and forts built by the colonialists, dating to the seventeenth and eighteenth centuries, form an integral part of the cultural tourism potential of the area. The area also possesses diverse flora and fauna, including some of the tallest trees in the world, a population of forest elephants and other endangered species including Bongos and Diana monkeys, and a 5-kilometre stretch of undeveloped beach with rolling hills and lagoons. The area possesses one of the few national parks in West Africa, the 'Kakum national park'.

To aid the conservation of these rich natural and cultural resources, the United Nations Development Programme (UNDP) donated US $3.4 million, which was matched by the Ghanian government. However, there still existed a shortfall of money to carry out the planned conservation of the area. Subsequently, two international NGOs, the Smithsonian Institute (SI) and Conservation International (CI), in co-operation with the Ghanian government, used the debt-for-nature scheme to raise the extra money. In effect this meant the NGOs paying US $250,000 to redeem a debt with a face value of US $1 million.

To pay for the continuing conservation of the area, and provide employment opportunities for local people, which will help to stop the poaching of the fauna for economic reasons, the scheme involves the development of tourism that is compatible with the fauna and flora of the area. Initiatives that have already begun include an audit of the resources of Kakum national park by CI, and technical assistance for the restoration of two of the most significant castles and one fort in the locality. Interpretive trails have been developed and canopy viewing platforms constructed in the forests. The plan also emphasises the need for community participation in shaping tourism, and a plan for the development of tourism on the beach has already been produced by local people, which includes extensive agricultural areas. Not only will these areas help to provide employment opportunities for local people but they will also strengthen links between the agricultural and tourism sector. Local people have also been given training as tour guides and camping areas are being developed in a controlled fashion to be run by local people.

Source: Brown (1998)

swaps'. The basis of this approach is that a conservation-based non-governmental organisation (NGO) buys off some of a country's national debt, in return for guarantees that the indebted country will look after the conservation of a designated area such as a national park. This will usually involve the development of appropriate management plans for the park. In effect the NGO is purchasing the 'development rights' or alternatively the 'no-use' rights to a particular area of a country. As Pearce (1993a) points out, debt-for-nature swaps are the only way in which estimates of the 'existence' (i.e. non-use) value of nature have been established.

According to Stone (1993), the concept of the 'debt for nature' swap is an extension of the proposal made by Jomo Kenyatta the founder of modern Kenya at a conference in 1961, when he suggested that if African wildlife was a world possession then 'the world could pay for it'. The process is based upon the buying of national debts on the world's money market at a fraction of their face value; for example, an NGO may pay to a bank 25 cents for every dollar of government debt. Banks are willing to discount the value of the loan because of the risk that the indebted country may default upon its payment completely. Debt-for-nature swaps usually involve the co-operation of national government, international and local NGOs, and banks. Essentially a government is giving up its sovereignty on a designated area of its country to an NGO. International NGOs, such as Conservation International, World Wide Fund for Nature, Nature Conservancy, and government agencies such as the US Agency for International Development (USAID), have encouraged less developed countries towards conservation efforts by devoting increased attention and funding to the preservation of habitats and conservation of flora and fauna (Brown, 1998). Debt-for-nature swaps have been used in Madagascar, where the Agency for International Development (AID) paid US $1 million to purchase a part of the government's debt in exchange for the support of local environmental groups; and in the Philippines and Zambia in conjunction with the World Wide Fund for Nature (Stone, 1993). An example of the workings of a debt-for-nature swap is shown in Box 4.7.

Summary

- The occurrence of environmental problems resulting from economic growth has led to a questioning of the efficiency of the free market system to account for the total social costs of production and consumption. Of particular concern is the overuse of common resources, both in the sense of exhausting supplies and the exceeding of their capability to assimilate the waste associated with production, leading to damage to ecosystems.

- Measures of society's progress are also being questioned. The use of economic indicators, such as Gross National Product, emphasises that progress can be statistically measured and that economic growth is a good thing. However, many of the things that add to our 'quality of life', such as clean air, beautiful views, family and spirituality, cannot be measured. Indeed economic growth may be harmful to these things. Success in tourism is also statistically measured by governments in terms of the numbers of international arrivals, total tourist expenditure, and world market share. However, such figures reveal little about the full environmental and social costs of tourism development.

- Being able to give an economic value to the environment can aid its conservation. Economic evaluation of the environment based upon its use for tourism can help stave off more environmentally threatening development options. The revenues from tourism can be used both to encourage the economic development of communities based in natural areas, and also provide revenues for the management of the area, in a sustainable fashion.

- Although advances are being made in economic methodologies to cost environmental goods into the market-place, many environmentalists would argue that nature has an intrinsic value, which is independent of humans' willingness to pay for it. Ethical arguments will remain about how humans use the environment, which cannot necessarily be resolved by costing environmental goods into the market mechanism.

Further reading

Cairncross, F. (1991) *Costing the Earth*, London: The Economist Books.

Pearce, D. (1993) *Economic Values and the Natural World*, London: Earthscan Publications.

Sinclair, T.M. and Stabler, M. (1997) *The Economics of Tourism*, London: Routledge.

5 The environmental planning and management of tourism

- Legislation for environmental protection
- Environmental audits and management systems
- Codes of conduct in tourism
- Roles of different stakeholders in the environmental planning and management of tourism

Introduction

The evident need for the environmental planning and management of tourism, in light of the sometimes negative interaction between tourism and the physical environment, has become of concern to governments, non-governmental organisations (NGOs), local communities and the private sector. All of these different stakeholders in tourism have a role to play in influencing the development of tourism and influencing the extent to which its interaction with the environment is positive or negative.

The role of government

Although historically the role of government in tourism has been concerned primarily with realising economic benefits, there is a growing awareness that these benefits cannot continue to be realised if the natural resource base is allowed to decline. Through the passing of legislation and use of fiscal controls, governments have potentially a wide range of powers that they can exert to control the scale and type of tourism development, and to pursue policies for wider care of the environment.

However, the prioritisation given by a government to environmental protection will be a reflection of its own philosophy upon the role of the environment in development. The typical prioritisation of the physical environment in a hierarchy of national goals is shown in Figure 5.1.

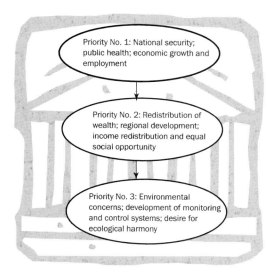

Priority No. 1: National security; public health; economic growth and employment

Priority No. 2: Redistribution of wealth; regional development; income redistribution and equal social opportunity

Priority No. 3: Environmental concerns; development of monitoring and control systems; desire for ecological harmony

Figure 5.1 *A hierarchy of national goals*

Source: After O'Riordan (1981)

O'Riordan's (1981) ordering of national priorities is reflective of the western view of development, in which the environment has been generally interpreted in a utilitarian fashion to be used to meet the needs and words of humans, and the process of economic growth has been separated from consideration of its effects upon the environment. Analysing O'Riordan's hierarchy, it is also evident that a country's stage of economic evolution is likely to influence the emphasis that is placed upon care of the environment. Therefore, for many less developed countries, tourism development policy is likely to emphasise the economic objectives of growth, such as employment creation and regional development, rather than how tourism and the environment can have a more symbiotic existence. Within this hierarchy, the environment is seemingly regarded as something of a luxury, to be prioritised after the achievement of other goals.

Although it can be argued that there is a higher level of awareness amongst policy makers, over the need to pay more attention to environmental care and stewardship than twenty years ago, there still exists conflict over the extent to which stewardship of the environment should detract from economic growth. An allegation made by many less developed countries against the western world, which seemingly wishes to dictate environmental policy on a global level, is that the West has the luxury to be able to be concerned over the environment because it has largely fulfilled many of the national goals identified by O'Riordan, whilst less developed countries are still struggling to fulfil them. As Stone (1993: 42) comments: 'Joan Martin Brown, an official with UNEP [author's note: United Nations Environmental Programme], has remarked: "The Third World says you're telling us not to do what you did to achieve your high standard of living. What are you going to do for us?"'

It is therefore likely that the extent to which governments are willing to pursue policies where the conservation of the environment is made a priority will depend greatly upon their level of economic development, and awareness about how the environment can be used in a sustainable fashion to generate wealth. Pressure to use the environment in an unrestricted fashion for wealth creation is exacerbated within the global economic environment. By encouraging the deregulation of trade and the liberalisation of economies, it is difficult for governments to refuse foreign investments, particularly in situations where countries are faced with a large variety of social and economic problems. Instrumental to the encouragement of this process have been international trade agreements, notably the GATT (General Agreement on Tariffs and Trade) and GATS (General Agreement on Trade in Services), which is specially relevant to tourism. These agreements are administered by the World Trade Organisation, which has been operational since 1995. A basic premise of GATS is free market access, and the denial of any protectionist measures, in effect opening up countries to unlimited amounts of foreign investments.

This increasing ability of transnational investors to switch capital to other countries means that many governments are likely to be reluctant to place any obstacles, such as environmental regulations, in the way of the inflow of capital for tourism development projects. For example, the demand of government X for mandatory environmental impact assessments on large-scale tourism development projects may mean costly delays to a foreign investor, inducing a longer payback period and a reduced rate of return on their investment. Subsequently, the investor may decide to place their investment into country Y, where environmental regulations are more lax or non-existent. Country X now loses the potential economic benefits from the project, and the government will also undoubtedly court political unpopularity for refusing the investment. The environment may consequently come to be viewed by some of the populace as a hindrance to development, rather than something to be cherished and forming the basis of their long-term prosperity.

Additionally, debt repayments to western banks and governments, combined in some cases with strict external controls on government expenditure associated with 'structural adjustment' of some national economies demanded by the International Monetary Fund (IMF), greatly influence the ability of countries to allocate resources for environmental protection. 'Structural adjustment' refers to a set of free market economic policies imposed by the World Bank and IMF as a condition of receiving financial help. Part of this adjustment involves reductions in government spending to reduce a country's debt level. Subsequently, if conservation of the environment is low in the list of government priorities, it may be one of the first areas to be cut back upon in government expenditure.

Certainly where a government favours no kind of regulation of commercial activity then the pace of development is likely to be quicker, but the consequences for the environment will be less easy to monitor, and are more likely to be destructive. An example of the potential consequences of inward investment for tourism with little government control is illustrated in the case study of Langkawi in Malaysia, as is described in Box 5.1.

Box 5.1

Uncontrolled inward investment for tourism development in Langkawi, Malaysia

In the 1980s, the Malaysian government decided that Malaysia needed to have a well-known destination to help promote its tourism. Subsequently, it decided to develop one of the archipelago of islands known as Langkawi, situated off the west coast of peninsular Malaysia. A decision was taken to develop a tourism complex to be named the 'Langkawi Resort', on 3,500 acres of land around a beautiful bay named Tanjung Rhu, at a total investment cost of US $1 billion in 1980s prices. This bay was specifically chosen because of its outstanding beauty, including casuarina trees, a beach front carpeted with white flowers, and its crystal clear lagoon and waterways. The development company was called 'Promet', the largest shareholder being Singaporean, and financial support was also given by the federal and state governments for infrastructure development, including an international airport. Initial construction work began in 1984, but by 1985, Promet was put into receivership. Today Tanjung Rhu is a devastated area resembling a wasteland or moonscape, with jungle and mangrove swamps having been completely cleared, the sand having been removed from the beach for use in construction and the water being silty and unclear.

Source: Bird (1989)

Although national priorities and global economic forces may detract from environmental conservation, governments do have a range of policy and legislative measures at their disposal which can be used to safeguard environmental resources, if they wish to use them. Typical measures that governments at national, regional and local levels can consider, include a mix of policy and planning measures. These include the following:

- The establishment of protected areas through legislation, for example national parks, and application for international recognition of significant environments, such as the World Heritage Site (WHS) status given by UNESCO;
- the implementation of land-use planning measures such as zoning, carrying capacity analysis, and limits of acceptable change (LAC) to control development;
- mandatory use of environmental impact analysis (EIA) for certain types of projects; and
- encouraging co-ordination between government departments over the implementation of environmental policy, and entering into dialogue with the private sector to encourage the adoption of environmental management policies, such as environmental auditing and the development of environmental management systems.

Protected areas

Governments possess the powers of legislation to establish protected areas and a variety of different types of designations exist. Based upon work carried out under the patronage of the United Nations Environmental Programme (UNEP), the World Tourism Organisation (1992) has produced several classifications of protected areas. Arranged by order of the degree of their permitted use by humans, the most restrictive being at the top and the less restrictive at the bottom, they are as follows:

- *scientific reserve/strict nature reserve* – the main purposes of this designation are to maintain and protect the existing ecological balance of the area for scientific study and for environmental education;
- *national parks* – to protect outstanding natural and scenic areas for educational, scientific and recreational use. Generally they tend to cover large areas of land, are not materially altered by human activity, and extractive industries are not permitted inside their boundaries;
- *natural monuments/ natural landmarks* – the aim of this designation is the protection and preservation of nationally significant natural features defined by their special interest or unique characteristics;
- *managed nature reserve/wildlife sanctuary* – human manipulation, for example the culling of a predatory species, is involved in these areas, to ensure the protection of nationally significant species of the biotic communities and physical features of the landscape;

- *protected landscapes* – emphasis is placed on the maintenance of nationally significant landscapes which are characteristic of the harmonious interaction of humans and nature. Emphasis is also placed upon the enjoyment of the area through recreation and tourism, as long as this does not detract from the normal lifestyle and economic activity of the area;
- *resource reserve* – the aim of this designation is to protect or sustain the resources of an area for future use by prohibiting development activities that form a threat to them;
- *naturally biotic area/anthropological reserve* – to permit the way of life for societies that are living in harmony with the environment to continue, uninterrupted by modern technology and human activity; and
- *multiple-use management area/managed resources* – this area can be used for the sustained production of a mix of water, timber, wildlife, pasture and outdoor recreation. The conservation of nature is orientated to the support of economic activities, although specific zones can also be designed within these areas to achieve specific conservation objectives.

For some of these protected area status tourism has direct relevance in terms of both benefiting from the protection of the environment from other forms of development and, if planned and managed appropriately, being able to make a positive economic contribution to environmental protection. For instance, carefully regulated and managed tourism, such as small groups interested in scientific education, could help fund the research and protection of a scientific reserve, whilst tourism revenues have already directly aided the establishment of national parks especially in less developed countries. By the early 1990s, protected areas covered nearly 5 per cent of the earth's surface, with over 130 countries having some area of their land surface under protected area status (World Tourism Organisation, 1992).

One of the most familiar protected area designations worldwide, and one in which tourism plays an important role, is the national park. The responsibility for passing legislation to establish a national park lies with governments. They are usually established with the objectives of protecting outstanding natural areas from overdevelopment and providing areas of access to nature for tourists and recreationists. The first national park in the world was created at Yosemite in America in 1890, and it is not an accident that its establishment coincided with the increasing urbanisation of America, in the latter half of the nineteenth century. The focus of the rationale for establishing Yosemite national park, as it was to be in the United Kingdom approximately sixty years later, was not based upon conservation but upon the provision of nature for the enjoyment of urban dwellers. Based upon the example of the rationale for creating the American and British national park systems, the creation of national parks can be interpreted as being anthropocentrically driven, placing emphasis on the conservation of nature for our own enjoyment rather than for its intrinsic value.

National parks may also play a greater part in the national psyche than just purely being areas for nature conservation and recreation. Hall and Lew (1998) suggest that the development of the Yosemite National Park, and the need for a rapidly urbanising American population to stay in contact with nature, was to act as a reminder of the pioneer mentality that modern America had been built around as much as for recreational purposes. An alternative perspective on the creation of national parks is also given by MacCannell (1992: 115):

> The great parks, even great urban parks, Golden Gate in San Francisco or Central in New York, but especially the National Parks, are symptomatic of guilt which accompanies the impulse to destroy nature. We destroy on an unprecedented scale, then in response to our wrongs, we create parks which re-stage the nature/society opposition now entirely framed by society.

In MacCannell's view, the creation of national parks is symptomatic of a release of national guilt over the harm humans have caused to the non-human environment, during the process of industrial modernisation. However, he also suggests that although the creation of the parks represents a 'good deed' of industrial civilisation, their creation also helps to affirm the power of humans over nature. The possible guilt experienced subconsciously by western society, over the way it has instrumentally used nature in the process of development, may also influence other aspects of perceptions of tourism. The fact that tourism often takes place in relatively unspoilt environments helps us to imagine how nature and life may have been before industrialisation and development took place. It may therefore be difficult for many in society to perceive that tourism can be environmentally threatening, as many of the environments it takes place in, certainly in the early stages of tourism development are the antithesis of what has gone wrong in developed societies.

Ironically, today, the biggest threat to many national parks in the developed world is tourism. For instance, the success of Yosemite in attracting tourists is without doubt, as Murphy comments (1985: 41):

> In addition to serving the drive-through visitor, Yosemite accommodates those who wish to stop-over, providing campsites, cabins, and a hotel in Yosemite Village. At times the valley which contains this village and attracts most of the visitors becomes an off-shoot of Los Angeles, complete with traffic jams and smog conditions – the very conditions tourists are trying to escape.

Overpopularity is perhaps one of the greatest hidden dangers to the environment that the parks are trying to protect, and by giving an area the status of a national park it automatically becomes an attractive place to visit. This can be especially problematic for park management in areas where national park status was given several decades ago, but the pace of technological change and transport development has meant that they have become easily accessible to large numbers of people for day trips and vacations.

Although many countries possess national parks, the type and amount of permitted human activity in them may vary considerably. The creation of national parks is not restricted purely to terrestrial areas of the world's surface, for instance marine parks have also been established to protect coral reefs in places such as the Netherlands Antilles (Bonaire Marine Park), and in the Seychelles (Saint Anne National Marine Parks). In the United Kingdom, the decision to create national parks resulted from the Addison Committee's enquiry into the feasibility of national parks and nature reserves in 1929. However, it was not until 1949 that the British government passed the 'National Parks and Access to the Countryside Act', creating the first national park. As is suggested in the name of the act, the rationale behind the creation of the British National Parks system was to provide a place for urban dwellers to have access to the countryside, as much as to offer protection to the environment from development. When the parks were created they already included human settlements and associated economic activities such as agriculture, forestry, and some extractive industries. Ultimately, the type of national park or protected area that is established in a country is likely to be a reflection of a variety of different factors, including existing levels of economic development, population density and the extent of institutional and financial support from government.

In less developed countries, the rationale for the establishment of national parks is much more closely associated with the conservation of wildlife supported by the revenues from tourism. Indeed, the day-to-day running and operation of the parks is often dependent upon the revenues received from international tourism. In less developed countries, national parks act as important focal points for attracting international tourists, for example as in east and south Africa, Costa Rica, India, Nepal and Indonesia (World Tourism Organisation, 1992). However, sometimes the creation of national parks in less developed countries has adverse cultural impacts, including the displacement of indigenous peoples, as was discussed in Chapter 3. Although national parks are very important for aiding conservation of the natural environment, their creation can result in costs besides benefits, as summarised in Box 5.2. The key to achieving success in national parks is the development and implementation of suitable management plans, which balance the use of natural resources, the needs of the local people and the expectations of the tourists.

Box 5.2

Summary of the costs and benefits of national parks

Benefit	Cost
Protect landscapes, wildlife and ecological communities	Unless carefully managed, recreation and tourism can pose a threat to both the landscape and wildlife that the park was established to protect
Provide a place for people to have access to and experience the countryside. Tourists can also provide revenues for scientific research and conservation projects	Granting of national park status focuses attention on the area. This may possibly lead to the attraction of too many tourists and overcrowding of the area
Offer employment opportunities for local people to become involved in conservation of the environment rather than destructive practices such as clearing natural vegetation for agriculture and poaching	Indigenous peoples can be excluded from their territory to protect landscape and wildlife

Besides legislating for the creation of national parks, other types of protected area status may be used by governments, as is exemplified in the case of the Balearic Isles in Box 5.3.

Box 5.3

The use of legislation for protected area status in the Balearic Isles

The growth of tourism in Spain since the unveiling of Franco's 'Plan Nacional de Estabilization', in 1959, established a policy of 'crecimiento al cualquier precio' (growth at any price). The pressure this policy has placed on the environment in the Balearic Isles of Spain, led the Balearic Isles' government to pass a series of laws in 1991, to limit new tourism construction in both the physical and urban environments. Under the new legislation one-third of Mallorca's surface area is now protected from future development, based upon protective legislation of three different types:

- Natural Areas of Special Interest – areas deemed to be of outstanding natural value and ecological importance;
- Rural Areas of Scenic Interest – includes areas of primarily traditional land-use activity, but that are still deemed of special scenic value;
- Areas of Settlement in a Landscape of Interest – includes areas of a primarily urban nature although declared of an exceptional scenic value.

In all these areas no new construction will be permitted, beyond that of the existing land-use and infrastructure requirements, for example urban regeneration projects.

Source: Gamero, E. (1992)

Land-use planning measures

Besides designating protected area status for environments, there are a variety of other land-use planning measures that can be encouraged by governments to mitigate the negative effects of tourism.

Zoning

Zoning is a land management strategy that can be applied on different spatial scales, for instance within a protected area, or at a regional and even national level. According to Williams (1998: 111):

> Spatial zoning is an established land management strategy that aims to integrate tourism into environments by defining areas of land that have differing suitabilities or capacities for tourism. Hence zoning of land may be used to exclude tourists from primary conservation areas; to focus environmentally abrasive activities into locations that have been specially prepared for such events; or to focus general visitors into a limited number of locations where their needs may be met and their impacts controlled and managed.

Zoning can provide a proper recognition of the resources that exist in the area and subsequently identify where tourism can and can not take place. With specific reference to the use of zoning in protected areas the WTO (1992: 26) remark: 'a protected area can be divided into zones of strict protection (a 'sanctuary zone', where people are excluded), wilderness (where visitors are permitted only on foot), tourism (where visitors are encouraged in various compatible ways), and development where facilities are concentrated'.

An example of how zoning has been used in an attempt to balance the requirements of scientific research, conservation, tourism and other forms of commercial activity is on the Great Barrier Reef Marine Park in Australia, as is described in Box 5.4.

Another example of land-use zoning takes place in the Canadian National Park system. The Canadian Parks Service manages thirty-four national parks, and one marine park, covering a total area of 180,000 square kilometres. Five zones have been designated for application in the national parks, categorised by the resource base of the area and the amount of recreational access that is allowed there, as follows:

- zone 1 – 'special preservation' – areas that contain strictly protected rare or endangered species and where access is strictly controlled;
- zone 2 – 'wilderness' – represents 60 to 90 per cent of the park area and the primary aim is resource preservation. Use is dispersed with only limited facilities;
- zone 3 – 'natural environment' – this area acts as a buffer zone between zones 2 and 4 and access is primarily non-motorised;
- zone 4 – 'recreation' – overnight facilities such as campsites are concentrated in this area; and
- zone 5 – 'park services' – this area is highly modified providing many services but represents less than 1 per cent of the park area.

Areas like zone 5 that are highly modified to suit the needs of certain types of tourists or recreationists, often encompassing a range of visitor

Box 5.4

Zoning in the Great Barrier Reef Marine Park, Australia

The Barrier Reef in Australia forms the world's longest coral reef, stretching for almost 2,000 kilometres along the north-eastern coast of Queensland. The reef is home to approximately 350 species of coral, 1,500 types of fish and 6 species of turtle. The development of international airports at the towns of Cairns and Townsville, which are conveniently situated for access to the reef, has meant that the number of tourists wanting to visit the reef has grown substantially since the late 1970s. The growth of tourism and other economic activities based upon the reef has meant that increased pressure is being placed upon it, and also that there is increased potential for conflicts between different user groups such as fishermen, tour operators (who can take groups of several hundred tourists on large catamarans and other boats to the reef), and recreationists such as scuba divers. The types of consequences that result from the use of the reef for tourism include: physical damage from anchors, moorings, snorkelling, diving and people walking on it; collecting of marine fauna; and the discharge of waste, litter and fuel.

The response to these problems was the establishment of the Great Barrier Reef Marine Park Authority (GBRMPA) to co-ordinate the management and development of the area. One of their functions was to zone the park to allow multi-use of the reef whilst preserving its ecology. They developed four different types of zones:

- *Preservation Zones* – areas in which use of the reef for virtually any purpose is prohibited;
- *Scientific Research Zones* – areas where scientific research is permitted under strict control;
- *Marine National Park Zones* – areas where scientific, educational and recreational uses are permitted; and
- *General Use Zones* – areas where some commercial and recreational fishing is permitted.

Commercial tourism is permitted in the last two zones, and the zoning process also includes the designation of Special Management Areas, in which reefs that are being intensively used for tourism or other purposes can be protected or conserved. Another aspect of the GBRMPA's role is the environmental impact management of the Reef. All proposed tourist operations are subjected to environmental assessment before they can be granted a permit to operate on the reef, and large-scale developments, or those that are assumed to produce unacceptable environmental impacts, have to prepare environmental impact statements.

Source: Simmons and Harris (1995)

attractions to persuade people to remain there instead of venturing into more environmentally vulnerable areas, are referred to by recreational planners as 'honeypots'. Besides being used within protected areas, honeypots can also be developed on a wider spatial scale, to stop tourism developing in areas of regions or countries where the nature has been identified as being significant. However, if communities are living in such areas, it is quite likely that people will complain if they feel they are missing out on the economic benefits and opportunities that are provided by tourism in other areas of the country because the environment in which they live has been identified as unique or special.

Carrying capacity analysis

One of the techniques most commonly referred to in the limited literature on tourism planning is that of 'carrying capacity analysis'. The origins of this concept can be traced to the last century, when concerns were being expressed over the levels of wildlife population that could be supported by the environment. In more recent times the concept has been applied to tourism and the World Tourism Organisation (1992: 23) define carrying capacity as being:

> fundamental to environmental protection and sustainable development. It refers to maximum use of any site without causing negative effects on the resources, reducing visitor satisfaction, or exerting adverse impact upon the society, economy and culture of the area. Carrying capacity limits can sometimes be difficult to quantify, but they are essential to planning for tourism and recreation.

Similarly, Mathieson and Wall (1982: 21) state: 'Carrying capacity is the maximum number of people who can use a site without an unacceptable alteration in the physical environment and without an unacceptable decline in the quality of the experience gained by the visitors.'

From these definitions it is evident that there are different elements to the concept of carrying capacity beyond physical considerations. According to Farrell (1992), there are at least four different types of carrying capacity, and O'Reilly (1986) identified economic, psychological, environmental and social carrying capacities as being of relevance to tourism. All have threshold levels beyond which the carrying capacity would be deemed to have been exceeded leading to a deterioration in the quality of the aspect under consideration. 'Economic carrying capacity' relates to the extent of the dependency of the economy upon tourism; 'psychological carrying capacity' is reflected in the expressed level of visitor satisfaction associated with the destination; 'environmental carrying capacity' is concerned with the extent and degree of impacts of tourism upon the physical environment; and 'social carrying capacity' is concerned with the reaction of the local community to tourism. The four carrying capacities are not independent of each other, but it may be possible to exceed the threshold limit of one capacity for a limited amount of time, without there being necessarily a detrimental effect upon another type of capacity. For example, it is possible that an increase in the number of walkers in a mountain area could lead to increased levels of destruction of flora from trampling, even threatening the ecological balance of the area, whilst the satisfaction of the visitors is not diminished. However, if the number of walkers continued to increase and damage to the environment increased proportionally, eventually the level of environmental damage would lead to a threshold level being crossed where it detracted from the level of satisfaction with the wilderness being experienced by the walkers. The whole notion of when damage occurs is debatable as Wight (1998: 78) remarks:

> The term damage refers to a change (an objective impact) and a value judgement that the impact exceeds some standard. It is best to keep these two separate. In terms of human impact, a certain number of hikers may lead to a certain amount of soil compaction. This is a change in the environment, but whether it is damage depends on management objectives, expert judgement and broader public values.

Early attempts in the field of tourism planning at identifying the carrying capacity of destination areas were preoccupied with trying to quantitatively determine the number of tourists that could be accommodated in an area, without causing 'unacceptable' environmental and social changes. Although the concept of carrying capacity had been evolving and developing in the field of recreation studies since the 1960s, it has only been since the late 1980s that the concept has become of interest to tourism researchers and planners. One notable exception to this observation was a study carried out for the Irish Tourist Board by the United Nations in 1966, which attempted to define the numbers of visitors that different destinations in Donegal in Ireland could tolerate, without harming the physical environment (Butler, 1997). However, owing to the highly complex nature of tourism, the notion of quantifying capacity limits is extremely problematical – not least because, as was discussed in Chapter 2, there are different types of tourist who will display different types of behaviour, which makes it difficult to legislate for the impacts they will have in a destination. A number of factors are likely to influence the carrying capacity of any particular destination, as shown in Box 5.5.

Box 5.5

Factors that will influence the carrying capacities of tourism destinations

- Fragility of the landscape to development and change
- Existing level of tourism development and supporting infrastructure, for example sewage treatment facilities
- The number of visitors
- The type of tourist and their behaviour
- The degree of emphasis placed on the environmental education of tourists and local people
- Economic divergence and dependency upon tourism
- Levels of unemployment and poverty
- Attitudes of local people to the environment and their willingness to exploit it for short-term gain
- The existing level of exposure of cultures and communities to outside influences and other lifestyles
- The level of organisation of destination management

Setting capacity limits across environmental, economic and social areas will inevitably involve value judgements. For example, it may be viewed as acceptable by decision makers to exceed cultural and environmental carrying capacity limits to maximise economic benefits, although the long-term sustainability of such a decision is questionable. Conversely, as is the situation in the kingdom of Bhutan where only a few thousand tourists per annum are permitted, to protect the physical and cultural environments of the country, the opportunity to maximise the economic benefits from tourism is missed. Political reasons may also make it difficult for fixed carrying capacities to be operationalised, as such an initiative would be reliant upon government intervention. Butler (1997) suggests that it would be politically unacceptable to the private sector for governments to intervene and regulate the capacity of destinations, as tourism represents a form of free enterprise of capitalism and competition, and the private sector is generally opposed to the external controls of government. Additionally, given the disparate nature of the resources used by tourists in destinations, and in some cases an absence or lack of clear ownership of these resources, the responsibility for their management is extremely problematic. This situation is akin to Hardin's (1968) 'Tragedy of the Commons' paper discussed in the last chapter.

Today the notion that there is a fixed ceiling, a threshold number of visitors which tourism development should not exceed, is largely discredited (World Tourism Organisation, 1992; Williams and Gill, 1994). Coccossis and Parpairis (1996: 160) comment:

> However, until our understanding of the interactions between the environment and development – human actions – is much more profound, the concept of carrying capacity cannot be used in planning and practice as an absolute tool offering exact measurements but, instead, as one which is under continuous revision, development and research.

Owing to the difficulty of quantification and fixed carrying capacity limits, more emphasis is being placed on indicator monitoring systems to identify potential problems, rather than trying to set absolute numerical limits of tourist numbers for destinations.

Limits of acceptable change

An evolution of the technique of carrying capacity is the 'limits of acceptable change' (LAC) or alternatively called the 'limits of acceptable use'. According to McCool (1996: 1): 'The Limits of Acceptable Change (LAC) planning system was developed in response to a growing recognition in the US that attempts to define and implement recreational carrying capacities for national park and wilderness protected areas were both excessively reductionist and failing.'

As is indicated in the above definition the LAC system, like carrying capacity, has its roots in wildlife management and recreational planning. It is only within the last few years that the technique has begun to be talked about within the context of tourism planning and its application to the field is at present very limited. The main deficiency of carrying capacity analysis, as pointed out in the preceding section, is that many of the problems associated with tourism are not necessarily a function of numbers but of people's behaviour. The advantage of the LAC system is that it does not attempt to quantify the numbers of tourists that can be accommodated in the area. Instead the premise of the LAC system is the specification of the acceptable environmental conditions of the area, incorporating social, economic and environmental dimensions, and also its potential for tourism (Wight, 1998). The system is therefore reliant upon identifying the desired social and environmental conditions in an area, which subsequently necessitates the involvement of the community in determining the desired conditions.

The mechanics of the LAC system involve the adoption of a set of indicators which are reflective of an area's environmental conditions, and against which standards and rates of change can be assessed. Typically, the indicators would relate to the state of the destination's natural resources, economic criteria, and the experiences of local people and tourists. The indicators would therefore be a mix of scientific and social measures. For example: the levels of water, air and noise pollution could be monitored; the percentage of the work force employed in the tourism sector assessed; crime rates and driving accidents associated with tourism recorded; and levels of tourist satisfaction evaluated. Such indicators would be symptomatic of the impact tourism is having within the destination, and the effect it is having on the quality of life of residents. The indicators are regularly monitored and evaluated, and strategies are identified by the managing authorities to rectify any problems, and to progress towards the desired environmental and social conditions that the LAC system is intended to help achieve. It is important to point out that, owing to the nature of the indicators, measurement cannot be purely scientific, but is also dependent upon a citizen input besides a professional one. As the name suggests, LAC accepts that some change is inevitable, and provides a framework to monitor that change.

Environmental impact analysis

Box 5.4 briefly referred to the role of environmental impact assessments (EIA) and environmental impact statements (EIS) within the context of the environmental management of the Great Barrier Reef in Australia. Environmental impact assessment, as is suggested in its name, is concerned with assessing the predicted effects of development upon the environment and thereby providing decision makers with information on the likely consequences of their decision to proceed with a development. The origins of EIAs as a formal part of planning procedure can be traced to the passing of the National Environmental Policy Act (NEPA) in the

USA in 1969, which required the preparation of EISs by federal agencies for all major projects. It is not a coincidence that the adoption of NEPA was also accompanied by increasing vocality from environmental groups over the effects of development upon the environment. Weston (1997: 5) comments: 'Indeed the introduction of EIA through the American NEPA was as much a response to political pressure from the growing environment lobby as it was an attempt to introduce a new planning technique.'

Since the passing of the NEPA, EIAs have become a widely discussed and used planning instrument to try to assess the likely consequences of development. The use of environmental assessment can vary from site-specific development to forming part of a strategic environmental assessment (SEA) aimed at examining the consequences of environmental policy. There is no set structure to the components of an EIA but it is generally assumed that EIAs would assess future levels of noise pollution, visual impact, air quality, hydrological impact, land-use and landscape changes associated with a development. Most EIAs involve five stages: identification of the impact; its measurement; interpretation of the significance of the impact; displaying the results of assessment; and the identification of appropriate monitoring schemes. Importantly, Weston (1997) draws attention to the point that within the EIA process there needs to be monitoring of the impacts during the construction/implementation and operational stages, and also auditing of the EIA process by comparing predicted and actual impacts. The different types of techniques of carrying out an EIA include the use of matrices, overlay analysis, adaptive environmental assessment and management (AEAM), and systems and network diagrams.

The types of tourism developments that would be subject to environmental impact assessments would include hotel complexes, visitor attractions, marinas, and associated infrastructure such as airports, roads, waste treatment and energy plants. The growing importance of tourism as a force of environmental change is recognised in the inclusion of tourism complexes in legislative requirements of the types of projects to be subjected to environmental impact assessment. For instance, since 1981 tourism complexes have been included with water and energy plants, industrial areas, ports, railways, roads and airports as requiring environmental impact assessments in the Republic of Korea, and all planned hotel or resort facilities with more than eighty rooms adjacent to environmentally sensitive areas such as rivers, coastal areas, lakes and beaches, or in the vicinity of national parks, are subject to EIAs in

Thailand (Wathern, 1988). For countries in the European Community, since the adoption in 1988 of the European Commission Directive (85/337/EEC) formulated in 1985, later updated in the 1997 EU Directive (97/11/EC), it is recommended that all ski-lifts, cable cars, roads, harbours, airfields, yacht marinas, holiday villages and hotel complexes are subjected to environmental impact assessment.

However, although EIAs may seem to be a good idea they do face a number of problems in their implementation. One major problem is the cost of preparation of an environmental impact statement, which will require a variety of specialists including geologists, hydrologists, geographers, environmental scientists, and possibly sociologists and anthropologists if the analysis is to incorporate a social impact statement (SIA). If the costs are to be borne by the developers then it may put some developers off going ahead with the scheme. If the EIA is paid for by the developer, there will be the question of its ownership and objectivity, which may bring its findings into disrepute. A further problem is predicting the timing of the impacts and when they are going to occur. Therefore it is necessary to distinguish between the impacts which will occur during construction, operation, and possible closure of the complex. Additionally, within the context of tourism development there is the additional problem that the majority of developments consists of small-scale enterprises which are not subjected to environmental impact assessment, and where the subsequent nature of environmental damage is incremental and cumulative (Butler, 1993). One method that attempts to deal with the problem of cumulative impacts is 'cumulative effects assessment' (CEA). This can refer to either the on-going effects of one particular project or the combined effects of a range of different projects. Certainly this technique would seem to have a direct relevance to tourism. However, it is as yet limited in its methodological approach, and there is a lack of information in established texts on environmental impact analysis about its application or implementation. A summary of the criticisms of EIAs is made in Box 5.6.

Government co-ordination

In addition to passing legislation for protected area status as discussed at the beginning of this chapter, governments also have an important role in co-ordinating different stakeholders' interests in tourism, and formulating a policy for tourism development. Given that tourism is diverse, and

---

Box 5.6

Criticisms of environmental impact analysis (EIA)

- Costly to implement because of its requirement for a variety of different specialists
- Problem of avoiding bias originating from the ownership of the study
- The preparation of the EIA can cause delays in the planning process and the process has been criticised for providing inadequate opportunities for public involvement
- Much tourism development is small scale, thereby falling out of mandatory requirements for an EIA which tend to be restricted to larger scale schemes. Owing to its small scale, environmental damage is incremental, cumulative and less easy to predict

---

involves a variety of interested stakeholders, it subsequently impinges upon a wide range of different government departments and ministries. Obvious ministerial interests in tourism relate to issues of finance and economics, transport, national security and environment. Sometimes the interests of the departments may be conflicting, for example increasing the numbers of tourists may be beneficial for the economy, but may have long-term negative environmental impacts which have to be dealt with by the Ministry of the Environment, and cause possible threats to security.

The complexity of tourism means that there is a necessity for cross-ministerial links between different departments if a successful tourism policy is to be developed. Traditional cross-ministerial links with tourism have typically been with economics and transport, but it is likely that ministries of the environment will have an increasing influence in shaping tourism policy in the future. Besides co-ordinating efforts between various ministries, governments also have a role to play in bringing together different sectors of the industry, and encouraging them to be more environmentally aware and responsible. An additional role of government in destinations where tourism is an integral part of the economy is to encourage the provision in schools of specific environmental education for tourism. Such programmes could also be modified and extended to raise the levels of awareness of issues concerning tourism and the environment amongst local adults and tourists.

One way the government can actively encourage more environmentally sustainable tourism development is through the use of fiscal measures. For decades many national governments have used financial incentives such as cheap loans, tax-free holidays and reduced energy tariffs to encourage the development of tourism facilities as was exemplified in Box 1.2. In a similar way financial incentives could be used to target and encourage the development of more environmentally sensitive forms of tourism, including the use of alternative technologies such as solar power. Taxes could also be levied on tourism and tourists, as discussed in Chapter 4, that are 'ring-fenced' for environmental improvements, bringing benefits to both local residents and tourists.

In addition to policy making and fiscal influences, land-planning regulations can also be used by governments at regional and local levels to control the density, type, and style of tourism development. Aesthetic 'pollution', caused by the adoption of a uniformity of building style for tourist accommodation, is a problem on developed areas of coastline as discussed in Chapter 3, and government at all levels can take steps to stop this occurring. Besides controlling the density and height of buildings, government can also encourage the use of local materials and handicraft skills, to reflect the culture of the area.

The role of the private sector

Although it is perhaps more encouraging to think of governments and the private sector in partnership, rather than as two unconnected entities, there are nevertheless initiatives that can be taken by the private sector to aid environmental quality that are independent of government. The extent to which the private sector decides to take the initiative with such measures will be a reflection of their wider environmental and social commitment.

As the major player in the tourism system the private sector has a major influence in determining the extent to which impacts from tourism upon the environment will be either positive or negative. In countries where tourism is interpreted by government as primarily a private sector activity that requires little regulation, a pro-active approach to environmental management from the private sector is of critical importance if the environmental resource base is to be conserved. Although it would seem private sector organisations, such as tour operators and hotel developers, are solely interested in maximising their profits, there are examples of where major companies operating in the field of tourism have taken initiatives to enhance their environmental operations.

Environmental auditing and environmental management systems

One type of environmental management technique that is relevant to tourism businesses is environmental auditing. Remarking about the process of environmental auditing Goodall (1994: 656) comments:

> Environmental auditing provides the basis for such business practice [improving the current environmental performance of tourism firms] and is consistent with the view of management as a controlled cyclic process based on continuous monitoring of impacts and change, the development of knowledge and the feeding back of these into decision making by formalised process.

The reasons why businesses may be encouraged to participate in environmental auditing probably fall into three main categories. First, the passing of environmental legislation and enforcement of punitive measures against tourism firms who are polluters of the environment may encourage a company to seek to improve their environmental quality. Second, if companies believe they can reduce their costs of operations and increase their profits through the utilisation of environmental auditing, they are likely to pursue auditing as a course of action. Third, some companies may be genuinely philanthropic and willing to adopt as many measures as they can reasonably afford, to benefit the physical and social environments.

According to Parviainen *et al.* (1995) an environmental or eco audit would cover aspects of environmental management, including the following: the company's environmental and purchasing policies; the adequacy of its communication of environmental practices to its staff and their level of environmental training; impacts of the business upon the surrounding physical environment, including features such as air, water, soil and ground water, noise and aesthetics; energy usage; and waste management and waste water schemes. They also point out that environmental audits form an integral part of a wider environmental management system (EMS) for businesses.

Environmental management systems integrate strategic objectives for the environmental quality of a company's operation with the practical aspects of environmental auditing. The first stage of an EMS is for a company to clearly state that they have an environmental commitment, which if taken seriously will subsequently influence the operations of the company. The next stage is to outline broad objectives of what they hope to achieve; for example, for a hotel one objective may be to reduce the amount of untreated waste emitted into the sea. The company would then carry out an eco-audit of their operations, determine realistic targets of what can be achieved within a certain time frame, and develop mechanisms to achieve the targets. An essential part of the scheme is the ongoing monitoring of operations to determine if the targets set for environmental improvements are being met. If they are not then strategies must be developed to rectify the situation. Developing an EMS is a long-term commitment and is likely to take several years, to incorporate all the different stages, from policy to review. The EMS system is not exclusive to any size of business but the resources available to any particular organisation will have an influence on the quality of the scheme. Importantly, it will require an investment of time and commitment from all employees of the organisation.

Within the EMS system, the eco-audit becomes a tool to evaluate the company's performance, and to make subsequent alterations to environmental policy and plans of action. The use of EMSs in the tourism industry is limited, yet it offers an approach for businesses that is both environmentally beneficial and proactive. The benefits to the industry of using EMSs include:

- the reduced risk of financial liability for environmental damage;
- improving customer relations;
- reducing operating costs;
- improving access to lenders, insurers and investors; and
- EMS's voluntary nature is an efficient way of improving environmental resources without regulatory requirements and government interference.

(After Todd and Williams (1996))

A scheme that demonstrates the advantages of a proactive stance by government towards influencing the relationship between tourism and the environment, and the advantages of a healthy public–private sector partnership, is one based upon environmental auditing that was encouraged by the Finnish government. The aim of the scheme was to encourage tourism businesses to undertake environmental audits to make themselves more aware of their environmental practices, and also to help to develop a more sustainable form of tourism in Finland. Importantly from the perspective of businesses, the environmental audits demonstrated that operational savings could be made by improving environmental practices, as is discussed in Box 5.7.

Box 5.7

Environmental auditing in Finland

In 1994 a pilot project was initiated by the Finnish Tourism Board to improve the environmental performance of tourism businesses in Finland. The scheme, known as the 'YSMEK' project, was funded by the Ministry of the Interior and the Ministry of Finance. YSMEK involved ten tourism organisations undertaking eco-audits. These included city hotels, farm tourism establishments, a ski resort, spa hotel, and training and exercise centres. The eco-audits concentrated primarily on the hotel and restaurant operations of the businesses, although for rural tourist enterprises, impacts upon agriculture and forestry were also included. As a consequence of the environmental audits, the companies have reduced their usage of disposable products, produce less waste and decreased their consumption of raw materials, water and energy. Average cost savings have been 10 to 15 per cent for electricity and 30 per cent for water.

Source: Parviainen *et al.* (1995)

One tourism company, that took the initiative over the impacts of its operations upon the environment at the beginning of the 1990s, is the large German based tour group Touristik Union International (TUI). The use of environmental audits has now become a regular part of TUI's business operations as described in Box 5.8.

In the United Kingdom, which has one of the largest outbound tourism markets that is reliant upon tour operators, there is also evidence that the environment is becoming more prominent in the larger tour operators' thinking and planning. For instance, the Thomson Travel Group has recently started to take environmental concerns more seriously in their business operations as is discussed in Box 5.9.

An example of a hotel chain that has taken the initiative to introduce an environmental management scheme into its operations is Grecotels, the largest hotel company in Greece. Initially, the scheme involved hotel auditing programmes, but the programme has now been extended to include the strengthening of links between Grecotels and local suppliers

Box 5.8

Touristik Union International (TUI)

Although the tour-operating industry can be criticised for generally having taken little interest in environmental concerns, one notable exception, operating in the mass tourism market, is the German tour group Touristik Union International (TUI). In terms of volumes of sales, TUI is one of the two major tourism companies in Europe (the other being the Thomson Travel Group who are United Kingdom based), selling holidays to approximately 5 million clients per annum. TUI's interest in the environment represents a mix of concern about the impacts of mass tourism and a pragmatic business sense to respond to the market demands of the German tourism market, where the environmental quality of destinations is known to have a critical influence on the level of customer satisfaction and subsequent demand. TUI has therefore realised that by investing in the protection and conservation of the environment, they are helping to safeguard their own financial success for the future. In 1990, TUI was the first tourism company to appoint an environmental manager as a member of its management board, and now has an established environmental unit in the company dealing specifically with environmental matters. Apart from carrying out environmental audits of TUI's operations, the unit also consults and liaises with the following: the governments of host countries; international and national public organisations holding responsibility for tourism and the environment; regional and local authorities; their business partners, including hotels, airlines, car rental companies; and, importantly, their customers, to make them aware of good environmental practices. The advantage of the TUI group in terms of influencing policy over the environment, is that owing to its size (nine tour operators, five hotel companies with 130 hotels, and 700 travel agencies) it has a huge political influence. The TUI approach is innovative in the tourism industry because it incorporates environmental protection as a fundamental management function in the organisation of its companies, thereby fulfilling a goal of Agenda 21, which is discussed in the next chapter.

in a way that is not detrimental to the environment, for example through the encouragement of organic agriculture to produce foodstuffs for use in the hotels, as discussed in Box 5.10.

An interesting aspect of the Grecotel programme, described in Box 5.10, is that it demonstrates how a major hotel player can strengthen linkages to other economic sectors in a sustainable way. In 1994, Grecotels appointed a professional economist specialising in organic farming to add technical support to Grecotel's 'green' agricultural activities, and in

Box 5.9

Environmental policy and the Thomson Travel Group

Thomson Travel Group is one of Europe's major travel groups housing a variety of operating companies, including tour operators, travel agents and an airline. In total it has in excess of 14,000 employees, 800 retail shops and an airline with a fleet of over 40 aircraft, with associated business operations in the UK, Ireland, Sweden, Norway, Denmark, Finland, Germany and Poland. Some of the best-known brand names in the British travel market belong to the group, including Thomson Holidays, Crystal International Travel Group, Lunn Poly and Britannia Airways. Approximately 4 million holidaymakers per annum travel with Thomson Holidays alone and at a group level, this increases to approximately 7 million. In 1998, the Thomson Travel Group floated on the London stock exchange to become a public limited company.

The size of the Thomson Travel Group means that it has substantial influence upon tourism development in many areas of the world. As such the group realises it has a responsibility to the environments and societies of destinations in which it operates and also to the consumer. Subsequently, environmental policy and management is becoming an increasingly important part of the group's operations. Fritidsresor, a Swedish based travel organisation, which was purchased by the group in 1998, was already established as one of the most environmentally aware tour operators – having established links with the World Wide Fund for Nature (WWF), and UNESCO World Heritage Sites.

The environmental policy of the Thomson Group covers different technical and social aspects of their operations. Emphasis is placed on making sure that all the aircraft operated by Britannia Airways meet the statutory requirements in terms of noise, aircraft emissions, fuel and noise. The raising of Thomson's employees' and customers' awareness of environmental issues is another important component of the policy. Social aspects of tourism are also included. For example, in certain destinations, Fritidsresor state in their contract that they will terminate a contract with any hotel supplier who is found to have involvement with child sex prostitution. In the future, it is likely that the group will explore ways of working more closely with local communities in destinations, and also encourage the development of environmental audit and management systems to cover all areas of their operations.

Box 5.10

Grecotels, environmental auditing and organic farming

Grecotels are the largest hotel company in Greece with 22 hotels and a bed capacity of approximately 11,000. In 1992 it was the first Mediterranean hotel group to form an environmental department, and initiated an environmental management and protection programme, which was initially supported by the European Commission. Between 1992 and 1993 professional eco-audits were conducted on the six hotels which at that time comprised the Grecotel chain. As a consequence of these audits, the following actions were taken, which are examples of the practices employed by the group in five different operational areas:

Technical matters
- improved biological treatment plants and re-use of water for garden irrigation whenever possible;
- solar-heated water, up to 100 per cent in some hotels;
- wherever possible, sea water is used in the swimming pools;
- the use of non-CFC fridges; and
- flow regulators on all bathroom taps to reduce water wastage.

Purchasing policy
- financed organic agricultural production in Crete to supply vegetables and organic produce to the Grecotels situated there;
- recycling and re-use of waste products wherever possible;
- toxic and dangerous chemicals replaced with more eco-friendly products or reduced usage; and
- promotion of Greek traditional local products in hotel shops and food outlets.

Built environment
- use of local building materials in new hotels and for the renovations of older Grecotels; and
- re-introduction of local plant species in hotel gardens.

Communication to personnel and guests
- training of all personnel in the environmental policy of Grecotels and environmental duties are included in all job descriptions;
- production and distribution of an eco-leaflet to all personnel encouraging ecological thinking in the home as well as the work place;
- results of programmes are given to guests and their active participation encouraged in schemes; and
- conveyance of the scheme and results to other hoteliers, tourism associations, tourism students and policy makers.

Cultural and social
- sponsorship of archaeological sites on Crete; and
- co-operation with environmental organisations to protect endangered species and their habitats, for example the Sea Turtle Protection in Rethymnon.

1995 the company funded an organic farming project. The use of environmental procedures, besides benefiting the environment, have also proved to be financially beneficial to Grecotels. For instance the use of garden flowering plants produced under organic principles has resulted in less of a requirement for fertiliser and pesticides, achieving a cost saving of US $52,000 in 1996. The use of hotel garden and kitchen organic waste to produce compost in the four hotels in the Rethymnon area of Crete has saved US $80,000. Links to other sectors of agriculture were also strengthened, including co-operation with Cretan vine growers in 1996 to produce 30,000 bottles of local wine for use in the hotels, which by 1997 had risen to 400,000 bottles of wine. In 1997, 96 per cent of the total wine consumption in hotels was produced by the wine-growers. An additional benefit of using local producers is that there is a saving in fuel consumption and reduction in air pollution due to the minimisation of transportation needs. Olive oil, herbs and other local products are also promoted in Grecotels, providing a means for local families to generate income.

Is it all good news?

The case studies from Finland, Germany, England and Greece illustrate the positive advantages for businesses from making environmental improvements in their procedures. Importantly, from a business perspective, the case studies demonstrate that through the use of environmental audits it is possible to reduce the operational costs of businesses. The role of government is also illustrated by the Finland case study as being important in aiding the implementation of auditing schemes in the sector. Government support is particularly important for the many small and medium-sized enterprises that constitute the bulk of the tourism industry, which may want to be involved in environmental improvement schemes, but lack the technical expertise to be able to do so. The Grecotel project also illustrates how a major player can be of benefit to other sectors of the economy, in terms of stimulating demand for local suppliers, whilst helping to improve the environmental standards of agriculture and quality of food for guests.

Although the examples included in this section of the chapter may make it seem that the tourism industry prioritises environmental responsibility, and that a sanguine view of the sector's future relationship with the environment can be taken, the reality is that the majority of tourism businesses remain loath to make a commitment to the environment. Unfortunately the examples of good environmental practice in the sector are relatively few. The attitude of much of the sector still remains that until customers demand environmental improvements, companies are unwilling to make environmental commitments, especially if they cost money and make inroads into their profit margins. In what is a very price-sensitive market, it would seem that until tourists demand more environmental probity from the sector or companies are forced to internalise the external costs of their operations, the commitment of the tourism industry to the environment will remain low. The willingness of governments to take action to regulate the sector environmentally would also seem to be lacking, which means that the emphasis for change is likely to rest with the consumer and the market, rather than through regulatory policy or private sector philanthropy.

Environmental codes of conduct for tourism

Voluntary codes of conduct, to mitigate the negative impacts of tourism and improve environmental quality, have been developed by government, the private sector, and non-governmental organisations (NGOs) in the last few years. The usefulness of codes of conduct in tourism *vis-à-vis* other approaches to improve tourism's interaction with the environment is described by the United Nations Environment Programme (UNEP) as follows:

> A wide range of instruments can be used to put the tourism industry on the path to sustainability. Regulations, of course, are – and will remain – essential for defining the legal framework within which the private sector should operate and for establishing minimum standards and processes. Economic instruments are also being increasingly used by governments to address environmental issues. However, voluntary proactive approaches are certainly the best way of ensuring long-term commitments and improvements.
>
> (UNEP, 1995: 3)

The primary aim of codes of conduct is to influence attitudes and modify behaviour (Mowforth and Munt, 1998). The objectives of codes of conduct for tourism are shown in Box 5.11.

These objectives cover a wide range of the stakeholders in tourism including the private sector, government, local communities and tourists. Consequently, the developers of codes of conduct come from a wide

---

Box 5.11

The objectives of codes of conduct for tourism

- Serve as a catalyst for dialogue between government agencies, industry sectors, community interests, environmental and cultural NGOs and other stakeholders in tourism development
- Create an awareness within industry and governments of the importance of sound environmental policies and management, and encourage them to promote a quality environment and therefore a sustainable industry
- Heighten awareness among international and domestic visitors of the importance of appropriate behaviour with respect to both the natural and cultural environment they experience
- Sensitise host populations to the importance of environmental protection and the host–guest relationship
- Encourage co-operation amongst industry sectors, government agencies, host communities and NGOs to achieve the goals listed above

Source: United Nations Environment Programme (1995: 8)

---

variety of organisations, including governments and national tourist boards, the tourism industry and trade associations, and non-governmental organisations, such as Tourism Concern and the World Wide Fund for Nature. Codes tend to be targeted at the tourism industry, local communities involved with tourism, and tourists. According to Goodall and Stabler (1997), the salient principles of the codes relating to the tourism industry include: the sustainable use of resources; reduction of environmental impacts, for example atmospheric emissions and the disposal of sewage; reducing waste and overconsumption, for example increasing the amount of recycling; showing sensitivity for wildlife and local culture; adopting internal environmental management strategies such as environmental auditing; support and involvement of the local economy by using local suppliers where possible; and pursuing responsible marketing. A typical example of a code of ethics developed by the industry is that of the Tourism Industry of Canada, as shown in Box 5.12.

Codes of conduct may also be established for local communities who are affected by tourism. Such codes can be helpful in the following ways: advising on the role of the local population in tourism development; safeguarding local cultures and traditions; educating the local population on the importance of maintaining a balance between conservation and

---

Box 5.12

Code of ethics for the tourism industry – the Canadian example

The Canadian Tourism Industry recognises the long-term sustainability of tourism in Canada depends on delivering a high quality product and a continuing welcoming spirit among our employees and within our host communities. It depends as well on the wise use and conservation of our natural resources; the protection and enhancement of our environment; and the preservation of our cultural, historic and aesthetic resources. Accordingly, in our policies, plans, decisions and actions, we will:

1. Commit to excellence in the quality of tourism and hospitality experiences provided to our clients through a motivated and caring staff.
2. Encourage an appreciation of, and respect for, our natural, cultural and aesthetic heritage among our clients, staff, and stakeholders, and within our communities.
3. Respect the values and aspirations of our host communities and strive to provide services and facilities in a manner which contributes to community identity, pride, aesthetics and the quality of life of residents.
4. Strive to achieve tourism development in a manner which harmonises economic objectives with the protection and enhancement of our natural, cultural and aesthetic heritage.
5. Be efficient in the use of all natural resources, manage waste in an environmentally responsible manner, and strive to eliminate or minimise pollution in all its forms.
6. Cooperate with our colleagues within the tourism industry and other industries, towards the goal of sustainable development and an improved quality of life for all Canadians.
7. Support tourists in their quest for a greater understanding and appreciation of nature and their neighbours in the global village. Work with and through national and international organisations to build a better world through tourism.

Source: Tourism Industry Association of Canada (1995)

---

economic development; and providing quality tourist products and experiences (United Nations Environment Programme, 1995). An example of a local community code is one developed by the non-governmental organisation, 'Tourism with Insight and Understanding', based in Germany. The code stresses the importance of community participation in tourism development, the need for respect of the local culture from tourists, and the need for tourism to play a balanced part in the economy of the region. An abridged version of the code is shown in Box 5.13.

Box 5.13

A code of conduct for host communities (abridged version)

1 Tourism should supplement our economy appropriately. We know however, that it also represents a danger to our culture and our environment. We therefore want to supervise and control its development so that our country may be preserved as a viable economic, social and natural environment.

2 By independent decision making in tourism development we mean that the host population should decide on and participate in all matters relevant to the development of their region; tourist development by, with and for the local population. We encourage many forms of community participation without neglecting the interests of minorities.

3 The tourism development we aim for is economically productive, socially responsible and environment-conscious. We are prepared to cease pursuing further development where it leads to an intolerable burden for our population and environment.

Source: United Nations Environment Programme (1995)

The final type of code is aimed specifically at the behaviour of tourists. Broad guidelines produced in such codes usually include the following: learning as much as possible about your destination; using suppliers (such as airlines, tour operators, travel agents and hotels) that demonstrate a commitment to environmental practices; respecting local cultures and traditions; aiding local conservation efforts; supporting the local economy by buying local goods and services; and using resources in an efficient and unwasteful manner. A typical code of conduct for tourist behaviour is shown in Box 5.14, produced by the non-governmental organisation, Tourism Concern, for trekkers in the Himalayas.

However, it is not necessarily the case that attempts to modify tourists' behaviour have to be through written codes. In an age of increasingly sophisticated media techniques and information technology, the use of visual imagery to raise tourists' levels of environmental awareness are important. An example of such a project, which also demonstrates how

Box 5.14

The Himalayan Tourist Code

By following these simple guidelines, you can help preserve the unique environment and ancient cultures of the Himalayas.

Protect the natural environment

- **Limit deforestation – make no open fires** and discourage others from doing so on your behalf. Where water is heated by scarce firewood, use as little as possible. When possible choose accommodation that uses kerosene or fuel efficient wood stoves.
- **Remove litter, burn or bury paper** and carry out all non-degradable litter. Graffiti are permanent examples of environment pollution.
- **Keep local water clean and avoid using pollutants** such as detergents in streams or springs. If no toilet facilities are available, make sure you are at least 30 metres away from water sources, and bury or cover wastes.
- **Plants should be left to flourish in their natural environment** – taking cuttings, seeds and roots is illegal in many parts of the Himalayas.
- **Help your guides and porters to follow conservation measures**.
- **When taking photographs, respect privacy** – ask permission and use restraint.
- **Respect Holy places** – preserve what you have come to see, never touch or remove religious objects. Shoes should be removed when visiting temples.
- **Giving to children encourages begging**. A donation to a project, health centre or school is a more constructive way to help.
- **You will be accepted and welcomed if you follow local customs**. Use only your right hand for eating and greeting. Do not share cutlery or cups, etc. It is polite to use both hands when giving or receiving gifts.
- **Respect for local etiquette earns you respect** – loose, light weight clothes are preferable to revealing shorts, skimpy tops and tight fitting action wear. Hand holding or kissing in public are disliked by local people.
- **Observe standard food and bed charges** but do not condone overcharging. Remember when you're shopping that the bargains you buy may only be possible because of low income to others.
- **Visitors who value local traditions encourage local pride and maintain local cultures**, please help local people gain a realistic view of life in Western Countries.

The Himalayas may change you –
please do not change them.
As a guest, respect local traditions,
protect local cultures, maintain local pride.
Be patient, friendly and sensitive
Remember – you are a guest

Source: Tourism Concern

partnerships between the different stakeholders in tourism can be environmentally beneficial, is the initiative taken by the Association of British Travel Agents (ABTA) with two United Kingdom non-governmental organisations, Voluntary Service Overseas (VSO) and Tourism Concern, to produce an in-flight video highlighting the social and cultural concerns associated with mass tourism in the Gambia. The video aims to raise the level of awareness amongst tourists about the physical and cultural environments of the Gambia, importantly exploring the needs and wishes of the Gambians themselves. It will be shown on flights to the Gambia operated by First Choice, a major UK mass tourism operator. Other videos are also expected to follow, about Thailand and Kenya, and also about general issues of tourism and the environment (Balmer, 2000).

Criticism of codes

Although the development of codes of conduct offers a way forward in making all the stakeholders in tourism aware of their environmental responsibilities, there have been a number of criticisms made of them. According to Mowforth and Munt there exist a number of problems and issues arising with codes, including (1998: 121): 'the monitoring and evaluation of codes of conduct; the conflict between codes as a form of marketing and codes as genuine attempts to improve the practice of tourism; regulation or voluntary self-regulation of the industry; and the variability between codes and the need for coordination'.

There is an obvious concern that the codes produced by industry represent little more than a cynical marketing ploy and an attempt to stave off any kind of government intervention in terms of environmental regulation. The production of a code by a tour operator printed in a brochure, for example, may satisfy a consumer who is searching for a more ethical holiday, but may mean relatively little in terms of changed business practices by the operator. There is also the major question of who, if anyone, will take responsibility for monitoring the effectiveness of the codes. This has meant there has been a lack of a general evaluation of codes as Mason and Mowforth (1996: 163) comment: 'There has been a clear lack of monitoring and evaluation of codes of conduct for the purpose of addressing their uptake and effectiveness.'

In similar fashion Goodall and Stabler (1997) talk of the limited practical usefulness of the codes because of their concentration upon principles, rather than informing tourist businesses on best environmental practice, and how this can be implemented in their own organisation. They also point out the spatial limitations of the vast majority of codes, which are destination based, and thus ignore the consequences of tourism in generating and transit areas.

Summary

- The role of government is critical in formulating environmental policy for tourism. Through legislative and fiscal powers it has a range of measures to encourage prioritisation of the environment by the different stakeholders involved in tourism. However, in what is essentially a sector of the economy that is heavily based upon free enterprise, any attempts at government regulation of the operations of the private sector are likely to be heavily resisted.

- Environmental auditing and management systems are key managerial techniques that can be adopted by the tourism industry. Not only can such techniques improve environmental quality, but they can also reduce companies' operating costs, besides improving their public image. However, for the many small and medium-sized enterprises that constitute the tourism industry, technical assistance will be needed to help them implement environmental improvements.

- Although some businesses in the tourism sector have been proactive in making environmental improvements to their operations, the majority remain reluctant to invest in environmental improvements. The lack of willingness of governments to regulate the sector environmentally means that the emphasis for environmental improvement is likely to rest with the consumer and the market, rather than through regulatory policy or private sector philanthropy.

- Codes of conduct can be developed and adapted by different stakeholders in tourism. For the private sector they can be interpreted as a form of self-regulation rather than regulation being imposed by government. However, criticisms can also be made of codes of conduct, including lack of monitoring and evaluation of their usefulness, and spatial limitations.

Further reading

Hall, C. and Lew, A. (eds) (1998) *Sustainable Tourism: A Geographical Perspective*, Harlow: Addison Wesley Longman.

Wathern, P. (ed.) (1988) *Environmental Impact Assessment: Theory and Practice*, London: Routledge.

6 Sustainability and tourism

- Origins of sustainable development
- Meaning of sustainable development
- Perspectives on its interpretation
- Its application to tourism

Introduction

Although improvements in economic methodology and approaches to include the environment as an integral part of the economic system, and the use of a variety of environmental planning and management techniques may aid environmental conservation, concern over the negative effects of development upon the environment has led to calls for a new conceptual approach to development. This conceptual approach is termed 'sustainable development', and at the beginning of the twenty-first century, it has become the new paradigm for all forms of development, including tourism.

The origins of 'sustainable development'

In contemporary terms the term 'sustainable development' is usually credited to the Brundtland Report, officially the report of the World Commission on Environment and Development (WCED, 1987). The origins of 'sustainability' as opposed to 'sustainable development' lie in concerns over conservation and can be traced back to the conservation movement of the mid-nineteenth century (Stabler and Goodall, 1996). The concept of 'sustainable development' first originated in the World Conservation Strategy published by the World Conservation Unit (IUCN) in 1980 (Reid, 1995).

However, the popularisation of term did not occur until its use in the Brundtland Report seven years later, perhaps because by 1987, environmental awareness and consciousness were at a much higher level than they had been seven years earlier. The report was based upon an inquiry into the state of the earth's environment, led by Gro Harlem Brundtland the Norwegian Prime Minister, at the request of the General Assembly of the United Nations. Concern over the effects of the pace of economic growth on the environment since the 1950s led the United Nations in 1984 to commission an independent group of twenty-two people from various member states representing both the developing and developed world, to identify long-term environmental strategies for the international community (Elliott, 1994).The key environmental concerns of the United Nations were the high levels of unsustainable resource usage associated with development, and the role of pollution in major environmental problems such as global warming and depletion of the ozone layer, which threatened human well-being.

Accompanying the heightened awareness of environmental problems was also a realisation that the environment and development are inexorably linked. Development cannot take place upon a deteriorating environmental resource base, neither can the environment be protected, when development excludes the costs of its destruction. Although 'development' is a common term in today's language, it is only since the 1950s that development has been studied as an academic subject, the time at which colonial territories started to achieve independence (Elliott, 1994). 'Development' and 'growth' are often used as synonymous terms but there is a critical difference between the two. Growth means to get bigger or larger, whilst development refers to a change in state, for the better. The difference between development and growth is aptly summed up in the following comment from Scottish National Heritage (1993: 9):

> To grow means to increase in size by the assimilation or accretion of materials. To develop means to expand or realise the potentialities of: to bring a fuller, greater, or better state. When something grows, it gets quantitatively bigger; when it develops, it gets qualitatively better, or at least different.

The pace of economic growth since the 1950s has been rapid; for example, although industrial production grew fiftyfold in the twentieth century, 80 per cent of this growth took place since 1950 (WCED, 1987). Whilst this rapid pace of industrial growth has helped to increase the living standards and life expectancy for many, the number of people in the world who continue to live in absolute poverty (i.e. having below sufficient income to meet the basic needs of food, shelter and clothing) has increased to 1 billion people or approximately 20 per cent of the world's population (Reid, 1995), although according to the *New Internationalist* (1999) the figure of one billion represents the same number as in the 1930s when records began. The way development has been pursued, characterised by a general lack of concern for the environment, has led to the use of natural resources in a way that is unsustainable, that is, many finite resources are being exhausted whilst the capacity of the natural environment to assimilate waste is being exceeded. To try to grasp an overall image of the global effects that have resulted from the pursuit of economic growth during the twentieth century is difficult, but examples of changes that have happened are given in Box 6.1.

Box 6.1

Changes related to development and the environment that took place in the twentieth century

- The world's population has increased from approximately 1,600 million in 1900 to approximately 6,000 million at the beginning of the twenty-first century
- Approximately 50 per cent of the world's rainforests have been destroyed post-1950
- Use of chlorofluorocarbons (CFCs), which are a major contributor to the breakdown of the ozone layer, rose from 100 tons in 1931 to 650,000 tons by the 1980s
- Over 50 per cent of the total increase in carbon dioxide levels between 1750 and 1990 occurred after 1950
- The proportion of people suffering from chronic malnutrition remains at over 1 billion, the same number as when records began in the 1930s. However, expressed as a percentage of the total world population the figure has fallen, although in Africa the proportion is on the increase again
- World consumption rose from US $1.5 trillion in 1900 to US $24 trillion in 1998, but the proportion of GNP per capita of low-income countries as a percentage of that in high-income countries, has reduced from 4 per cent in 1950 to 1.4 per cent in 1997
- Life expectancy has risen in developing countries from 25 years in 1900 to 61 in 2000

Source: *New Internationalist* (1999)

It was the predominance of the negative aspects of these changes that led to the calls for sustainable development. The term gained greater attention following the United Nations Conference on Environment and Development (UNCED), held in Rio de Janeiro in June 1992, popularly referred to as the 'Earth Summit'. At the Earth Summit a programme for promoting sustainable development throughout the world, known as Agenda 21, was adopted by participating countries. Agenda 21 is an action plan laying out the basic principles required to progress towards sustainability. It envisages national sustainable development strategies, involving local communities and people in a 'bottom-up' approach to development, rather than the 'top-down' approach which has typically characterised national development plans.

Although tourism as an economic sector was not debated in Rio, five years later at the 'Earth Summit II' in New York, it was debated as a recognised economic sector. In the recommendations and outcomes of the report it was stated:

> The expected growth in the tourism sector and the increasing reliance of many developing countries, including small island developing States, on this sector as a major employer and contributor to local, national, subregional and regional economies highlights the need for special attention to the relationship between environmental conservation and protection and sustainable tourism.
>
> (Osborn and Bigg, 1998: 169)

In the last decade of the twentieth century, the term 'sustainable development' became widely used by governments, international lending agencies, non-governmental organisations, the private sector and academia. An encapsulated history of the concept of sustainable development is displayed in Box 6.2.

The meaning of sustainable development

The fact that the term 'sustainable development' can be adopted by governments, international lending agencies, non-governmental organisations, the private sector and academia, some of whom could be viewed as having divergent and politically opposed objectives, is a reflection of the inherent ambiguity of the concept. This ambiguity permits a variety of perspectives to be taken on sustainability. Much of this ambiguity can be traced to the most commonly quoted definition of sustainable development taken from the Brundtland Report:

Box 6.2

The origins of sustainable development

- Term can be traced to the conservation movements of the mid-nineteenth century
- Established as a policy consideration in the World Conservation Strategy published by the World Conservation Union (IUCN) in 1980
- Term gains greater attention and popularity after the publication of the Brundtland Report (1987)
- 'Earth Summit' 1992, Rio de Janeiro, adopts 'Agenda 21' aimed at promoting sustainable development throughout the world
- Tourism is recognised as an economic sector that needs to develop sustainably at 'Earth Summit II' in 1997 at New York

> Yet in the end, sustainable development is not a fixed state of harmony, but rather a process of change in which the exploitation of resources, the direction of the investments, the orientation of technological development, and institutional change are made consistent with future as well as present needs.
>
> (WCED, 1987: 9)

Richardson (1997) describes this definition as a political fudge, aimed at compromising the opposing views of commissioners from different states, to keep everyone happy. However, the remainder of the Brundtland Report makes it clear that within the scope of this definition other key issues relating to development have to be addressed, such as the alleviation of poverty, degradation of the environment, and issues of intra- and inter-generational equity.

It is important to realise that sustainable development is not concerned with the preservation of the physical environment but with its development based upon sustainable principles. The Brundtland Report stresses the need for the alleviation of global poverty, not only as an ethical objective, but also as a key method to ameliorate the pressures being placed upon the physical environment. Subsequently the report lays out the principles of how the environment should be developed:

> Economic growth and development obviously involve changes in the physical ecosystem. Every ecosystem everywhere cannot be preserved intact. . . . In general, renewable resources like forests and fish stocks need not be depleted provided the rate of use is within the limits of regeneration and natural growth.
>
> (WCED, 1987: 45)

Emphasis is therefore placed upon the conservation of the resource base rather than the preservation of individual flora and fauna. In terms of the use of finite resources, such as minerals and fossil fuels, actions such as resource management, recycling and economy of use are highlighted to make sure that the resources do not run out before substitutes are found.

A central theme of the Brundtland Report, poverty alleviation through sustainable development, is critical for the long-term environmental well-being of the planet. Poverty is a major cause of environmental destruction, a relationship that is particularly exacerbated in regions of the world where the population is growing rapidly, and forced into more marginal environments. The link between poverty and environmental destruction is emphasised by Elliot (1994: 1) in the following passage:

> In the developing world, conditions such as rising poverty and mounting debt form the context in which individuals struggle to meet their basic needs for survival and nations wrestle to provide for their population. The outcome is often the destruction of the very resources with which such needs will have to be met in the future.

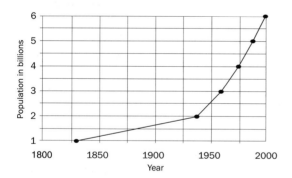

Figure 6.1 *World population growth, 1830–1998*

Poverty has been worsened in many areas by population pressure and the growth in world population in the twentieth century was rapid. In 1900 the world's population was 1,600 million, by the beginning of the new millennium it has risen to 6,000 million, and according to the United Nations by the year 2025 it is estimated it will be over 8,500 million (World Guide, 1997). The growth of world population is shown in Figure 6.1.

Different perspectives on sustainable development

That there should be differing perspectives on the meaning of sustainability is perhaps unsurprising, given the range of priorities, interests, beliefs and philosophies underpinning human interaction with the environment. Two broad ideological approaches to the environment are recognisable, 'technocentrism' and 'ecocentrism'. Technocentrism is characterised by a belief that technical solutions can be found to environmental problems through the application of science, thereby placing its faith in quantifiable solutions to problems. Its reliance upon quantification allows for a divorced objectivity in decision making, rendering subjective considerations over the environment, such as feelings or emotions, as being unworthy of consideration. Such a reliance upon objective measurement means that the environment's complexity as a system can be overlooked and differing viewpoints ignored. As Reid (1995: 131) remarks: 'Technocentrism's readiness to quantify problems and solutions and its forecasts of success make it attractive to decision-makers who may be ignorant of the complexities of issues filtered through viewpoints, interpretations and value judgements of which they may be unaware.'

Through a lack of appreciation of the complexity of the environment as a system, there is also a lack of realisation that most development decisions have consequences for the environment that extend beyond the physical boundaries of an individual project. Within the philosophy of technocentrism the physical environment is viewed as a resource to be exploited by humans as they deem appropriate. As for the degree of democracy in decision making, technocentrism is characterised by a preference for centralised control rather than local decision making, as Pepper (1993: 34) writes: 'There is little desire for genuine public participation in decision making, especially to the right of this ideology, or for debates about values.'

An alternative ideological approach to how we view the environment is ecocentrism (O'Riordan, 1981). Ecocentrism is closely associated with the philosophies of the romantic transcendentalists and characterised by a belief in the wonderment of nature. Ecocentrics have a lack of faith in both modern technology and technical and bureaucratic elites, and subsequently advocate alternative technologies as a way forward. This is not only because alternative technologies are likely to be more environmentally benign, but also because they are more democratic in the sense that they can be owned, maintained and understood by individuals with little economic or political power. Ecocentrics' position on technology may be termed 'Luddite', that is, they are not opposed to new technology, but are opposed to technology that places its ownership and control in the hands of a powerful elite.

Doyle and McEachern (1998) equate the term 'ecocentrism' with 'deep ecology', recognising that it has four main characteristics. These are as follows: that all beings, whether human or non-human, possess an intrinsic value, the antithesis of the technocentric instrumental viewpoint of nature; that all beings are of equal value and there is therefore no hierarchy of species in nature; that all nature is interconnected, with no dividing lines between the living and the non-living, the animate and inanimate, or the human and non-human; and that the earth is finite in its carrying capacity. These views are in contrast to the 'dominant world-view' on development, which incorporates a pioneering and frontier mentality, assuming an unlimited supply of natural resources and an unlimited waste absorption capability in nature. The differences over approaches to development between the 'dominant world view' and 'deep ecology' are shown in Box 6.3.

Box 6.3

Differences in views of development between the 'dominant world-view' and 'deep ecology'

Dominant world-view	Deep ecology
• Strong belief in technology for progress and solutions	• Favours low-scale technology that is self-reliant
• Natural world is valued as a resource rather than possessing intrinsic value	• Sense of wonder, reverence and moral obligation to the natural world
• Believes in ample resource reserves	• Recognises the 'rights' of nature are independent of humans
• Favours the objective and quantitative	• Recognises the subjective such as feelings and ethics
• Centralisation of power	• Favours local communities and localised decision making
• Encourages consumerism	• Encourages the use of appropriate technology
	• Recognises that the earth's resources are limited

Source: After Bartelmus (1994)

The equating of the dominant world view with the attitudes of technocentrism is not difficult and according to O'Riordan (1981: 1) the balance of power in decision making in society lies with the technocentrics: 'Technocentrists tend to be politically influential for they usually move in the same circles as the politically and economically powerful, who are soothed by the confidence of technocentric ideology and impressed by its presumption of knowledge.'

Within these two broad ideological approaches a range of contrasting positions can be taken on sustainable development. Baker *et al.* (1997) conceptualise a ladder of sustainable development moving upwards from a technocentric approach at the bottom to an ecocentric position at the top. At the bottom of the ladder is what they term the 'Treadmill' approach, focused on accruing material products and pursuing wealth creation. This approach is strongly technocentric in character, with little concern being displayed for the environment, beyond believing that human ingenuity can solve any environmental problem created by development. The second rung of the ladder is occupied by 'weak sustainable development', in which the aim 'is to integrate capitalist growth with environmental concerns' (Baker *et al.*, 1997: 13). In this perspective emphasis is placed upon continued economic growth, but the environmental costs of development are to be reflected in economic calculations and accounting procedures. Unlike the 'treadmill' approach, 'weak sustainable development' recognises the finite nature of many natural resources, and also recognises that the environment's capacity to assimilate waste is not limitless. Emphasis is therefore placed upon furthering the development of economics as a discipline to permit a better evaluation of environmental assets. A criticism of this approach is that it limits the value of the environment to that which can be quantitatively measured, neglecting its cultural or spiritual worth.

The third rung of the ladder is occupied by an approach more synonymous with ecocentrism than technocentrism. 'Strong sustainable development' advocates that environmental protection be a precondition of economic development. Subsequently, within this perspective the environment becomes the key consideration, rather than economic growth as in the latter two scenarios. This perspective requires that development policies aim to maintain the productive capacity of environmental assets, and preserve other environmental assets deemed worthy of protection as they are, for example tropical rainforests. 'Strong sustainable development' takes qualitative aspects of the environment into account and 'local communities' will be involved in decision making over development issues. All the available instruments of policy, including legal, fiscal and economic measures, should be used and adapted to support this approach.

Occupying the top rung of the ladder is what Baker refers to as the 'Ideal Model'. This approach is underpinned by a strong ethical dimension, the viewpoint being taken that nature and non-human life have an intrinsic value, which extends beyond their usefulness to humans. The measurement of growth expressed in quantitative terms ceases to be relevant as the 'quality of life' becomes the objective of development rather than the 'standard of living'.The policy implications of this viewpoint are that environmental protection will place severe restraints on the use of the earth's resources and economic activities. This approach represents a 'radical' approach to sustainable development, involving structural changes in global society and economy. Indeed, according to Baker, ecologists would argue that the 'Ideal Model' represents a new paradigm in its own right.

The advocation of radical changes in society to achieve sustainable development concentrates on power relationships in the wider political economy. This entails addressing the root causes of non-sustainability, including the distribution of power and wealth, the roles of transnational corporations, class-based politics, and gender inequalities. The distribution of intra-generational equity, even though it was an issue raised in the Brundtland Report as having direct consequences for poverty, is notably absent from most government agendas. Radical approaches to sustainable development challenge the values and principles of capitalist society, as Doyle and McEachern (1998: 37) comment: 'Radical environmental political theorists are involved in paradigm struggles, each seeking to create new sets of key values and principles that directly challenge existing, powerful paradigms.'

Radical political positions over the interpretation of sustainable development, that emphasise different degrees and elements of structural change include 'deep ecology', 'social-ecology', 'eco-socialism' and 'eco-feminism'. The values of *deep ecology* in contrast to the values of the dominant world view have already been illustrated in Box 6.3. To reiterate, the emphasis of deep ecology rests on the intrinsic value of all life, and acknowledging that all life is equal. This means that deep ecologists take the viewpoint that human beings are not the sole purpose of the evolutionary process but just another animal species. In this philosophy there is no theoretical separation between humans and the rest of nature. These beliefs have led to campaigns for the preservation of wilderness in North America, Scandinavia and Australia. Subsequently, deep ecologists are sometimes enthusiastic supporters of population control programmes in order to reduce destructive human pressures upon the earth, which has led to accusations of them holding fascist tendencies.

The politics of *social ecology* are rooted in anarchist traditions, with their associated hostility to the state, liberals and Marxists (Doyle and McEachern, 1998). Social ecologists subsequently advocate maximum individual autonomy and the decentralisation of society into local communities as opposed to the concept of the nation-state. According to Roussopoulos (1993: 122), social ecology: 'represents the greatest advance in twentieth century eco-philosophy'. Its development is closely associated with the American ecologist Murray Bookchin who adopts two key positions in his philosophical approach to social ecology. The first is to recognise that the many types of hierarchical structures that exist in society are socially constructed and determined, and represent the source of all forms of domination within society, and also between humans and nature. The second position, which distinguishes social ecology from deep ecology, is that although humans are part of nature they occupy a rather more exalted place in the evolutionary scheme (Doyle and McEachern, 1998).

Eco-socialists in contrast to social ecologists, attempt to integrate ecological principles with the political theories of socialism including Marxism, rather than anarchism. The concern of eco-socialism lies equally with the problems of social justice, such as the inequity of wealth distribution in society, and ecological concerns. Just as there are many types of socialism there are many varieties of eco-socialism. Although eco-socialism can be allied to Marxism, the theories of Marx that assume a limitless abundance of nature are not accepted by eco-socialists (Roussopoulos, 1993). A distinguishing feature between social ecologists and eco-socialists is their belief in the nation-state. Eco-socialists do not accept that environmental problems can be tackled at a localised level, favouring central government and the strengthening of pan-national bodies, like the United Nations. Consequently, a necessary precursor for environmental protection is for eco-socialists to have power.

According to Roussopoulos (1993) *eco-feminism* has its origins in anti-militarism and links the environmental problems that the world is facing today to the system of patriarchy. Therefore, ecological destruction and the oppression of women are linked. As for eco-socialism there is a broad range of differing political perspectives, including cultural eco-feminism, liberal eco-feminism, social eco-feminism, and socialist eco-feminism.

Political tension over sustainable development therefore underlies much of the debate about its interpretation. A fundamental division is between those for whom sustainability represents little more than improving technology and environmental accounting systems, whilst preserving the *status quo* of existing hierarchies and power structures in society, and those who have more radical political agendas involving changing the value systems and power structures of society.

Sustainability and tourism

In terms of the application of the concept of sustainable development to tourism, varying perspectives have been adopted, which are a reflection of the wider political debate about sustainable development outlined in the preceding section of this chapter. A broad differentiation can be made between 'sustainable tourism', in which emphasis is placed on the customer and marketing considerations of tourism to sustain the tourism sector, and using tourism as a vehicle to achieve 'sustainable development', in which emphasis is placed on developing tourism as a means to achieve wider social and environmental goals. Therefore 'sustainable tourism' will not necessarily equate with the aims and objectives of sustainable development.

Both Coccossis and Parpairis (1996) and Hunter (1996) identify a form of sustainable tourism which is oriented toward the viability of the tourism industry. Coccossis refers to this as the 'economic sustainability of tourism' and Hunter as the 'tourism imperative'. The aim of development is primarily concerned with satisfying the needs of tourists and players in the industry. The central justification of this approach is that tourism development is to be encouraged and seen as acceptable, if developing other economic sectors, such as logging and mining, are viewed as being more environmentally damaging than tourism. However, this perspective fails to allow for the fact that the negative consequences of tourism development are incremental and cumulative, and that the prejudgement of the likely environmental impacts of tourism development through techniques such as environmental impact analysis are therefore difficult, as was discussed in Chapter 5. Comparative impact analysis with other types of development alternatives therefore becomes meaningless and misleading.

Hunter's second position is when the environmental resources for tourism receive consideration, but are secondary to the growth of the tourism sector, an approach known as 'product-led tourism'. Hence although environmental and social concerns may be more important than in the former scenario, their relevance is still largely equated with maintaining the existing tourism product. Hunter suggests this is a 'weak' interpretation of sustainable development, but that such a position is justified in communities already heavily economically dependent upon tourism, where placing a high priority on environmental concerns may mean the well-being of the community is threatened. Development may have already led to extensive alteration of the natural environment and attention will therefore focus on environmental improvements to the existing development.

Hunter's third scenario is termed 'environmentally led tourism', in which types of tourism would be promoted that are reliant upon a high-quality environment. The main aim would be to make the link between the success of the tourism industry and environmental conservation so obvious to all the stakeholders that stewardship of the environment is a priority. This is analogous to what Coccossis refers to as 'sustainable tourism development', where the protection of the environment is seen as a key component of the long-term economic viability of the industry. The main difference compared to 'product-led tourism' is that the environment is prioritised and forms of tourism are developed that are not damaging to it. Tourism would therefore be centred upon attracting tourists who would wish to be educated about the natural environment and perhaps participate in its conservation.

The final scenario suggested by Hunter is 'neotenous tourism'. In some exceptionally sensitive areas, where the preservation of species is paramount because of their ecological significance, tourism should not be permitted at all. In other areas which are viewed as being ecologically significant tourism should be limited to very small numbers through policy control measures.

The focus of both Coccossis and Parpairis's (1996) and Hunter's (1996) work is the physical environment. However, the concept of sustainability should be broadened to include cultural, political and economic dimensions. The ambiguousness of the concept of sustainability means that the political context, and especially the political values of those who have power and decision making, will be influential in determining the interpretation of sustainable tourism. Butler (1998) points out that it is not possible to separate sustainable tourism from the value systems of those involved and the societies in which they exist. Concerning this latter point, Mowforth and Munt (1998: 122) remark: 'If it remains a "buzzword" which can be so widely interpreted that people of very different outlooks on a given issue can use it to support their cause, then it will suffer the same distortions to which older-established words such as "freedom" and "democracy" are subjected.'

Mowforth and Munt continue to demand that if sustainable tourism is really to be achieved then there is a 'need to politicise the tourism industry in order to promote its movement towards sustainability and away from its tendency to dominate, corrupt and transform nature, culture and society' (1998: 123).

House (1997) also infers a political dimension to the application of sustainability in tourism. She recognises two polarised positions or 'schools of thought' reflecting different ideologies on the employment of the concept. At one end are the 'reformists', whose ideology and actions of implementation are preoccupied with the *status quo*, whilst 'structuralists' possess a much more radical view of tourism development, which challenges the paradigm on which economic, social and political development is based. Reformists are therefore 'reluctant to challenge the existing social, political and economic structures that underpin tourism development and behaviour' (1997: 93).

Reformists do not challenge the values in society that create environmental and social problems but try to manage them. House (1997) draws a distinction between those who modify existing models of tourism to make them more 'sustainable', and those who use tourism as a vehicle to experiment with 'alternative existence approaches' to achieve an alternative social order to the present one. The structuralist model of tourism is therefore one which is radical, involving questioning the values of society, as much as those of tourism. The relationship between the ideas of 'reformists' and 'structuralists' and sustainable tourism is analogous with the differing positions taken on the meaning of sustainability within the field of political ecology, explained earlier in the chapter.

It is therefore necessary to realise that sustainable tourism is not merely connected with conservation or preservation of the physical environment but incorporates cultural, economic, and political dimensions. As has been pointed out, the ambiguity of the term means that it can be interpreted and owned by so many different groups with opposing ideologies that trying to agree a common definition of the term is meaningless. Perhaps the most useful way of thinking about

THINK POINT
Does 'sustainable tourism' mean the same thing as 'sustainable development'?

sustainability is not necessarily to think of it as an end point, but to think of it more as a guiding philosophy which incorporates certain principles concerning our interaction with the environment. The next section of this chapter considers the guiding principles that have been developed in connection with tourism.

Sustainability in action

The concept of sustainability has been applied in the tourism sector in different ways, at both national and local levels, and in the public and private sectors. The last decade of the twentieth century saw an increased effort by some private sector tourism organisations to make it evident that they were placing the environment in a more central position to their operations and attempting to become more 'sustainable', as was discussed in Chapter 5. The extent to which this is from a genuine concern for the environment, or a business ploy to attract more customers and an attempt to stave off regulation of the industry, is unsure. Butler (1998: 27) suggests that the tourism industry has adopted sustainability for three reasons: 'economics, public relations and marketing'.

One of the first public strategies on tourism and sustainability emerged from the Globe '90 conference in Canada, which brought together government, non-governmental organisations (NGOs), the tourism industry, and academics to discuss the future relationship of tourism with the environment. Five main goals of sustainable tourism were identified: '(1) to develop greater awareness and understanding of the significant contributions that tourism can make to the environment and economy; (2) to promote equity and development; (3) to improve the quality of life of the host community; (4) to provide a high quality of experience for the visitor; and (5) to maintain the quality of the environment on which the foregoing objectives depend' (Fennell, 1999: 14). As with the concept of sustainability, the goals tend to be all encompassing, potentially conflicting and give little guidance on how tourism should be developed.

In Britain, the Department of the Environment (DOE) developed guiding principles for the development of tourism in the early 1990s, as are shown in Box 6.4.

Box 6.4

Guiding principles of sustainable tourism

- The environment has an intrinsic value which outweighs its value as a tourism asset. Its enjoyment by future generations and its long-term survival must not be prejudiced by short-term considerations
- Tourism should be recognised as a positive factor with the potential to benefit the community and the place as well as the visitor
- The relationship between tourism and the environment must be managed so that the environment is sustainable in the long-term. Tourism must not be allowed to damage the resource, prejudice its future enjoyment or bring unacceptable impacts
- Tourism activities and development should respect the scale, nature and character of the place in which they are sited
- In any location, harmony must be sought between the needs of the visitor, the place and the host community
- In a dynamic world some change is inevitable and change can often be beneficial. Adaptation to change, however, should not be at the expense of any of these principles
- The tourism industry, local authorities and environmental agencies all have a duty to respect the above principles and to work together to achieve their practical realisation

Source: Department of the Environment (1991: 15)

The first of these principles is significant because of its recognition that the environment has an intrinsic value, that is, that nature has a consciousness and value in its own right as was discussed in Chapter 2. Notably, the recognition that this intrinsic value outweighs its value as a tourism asset may mean that the environment is excluded for use for tourism. The first principle also endorses the recommendations of the Brundtland Report by stressing the need for long-term planning considerations as opposed to short-termism.

The principles also stress other elements that are viewed as being essential to achieving sustainable development. The need for a balanced approach to development, respecting the character and nature of the place, is implicit to the principles. This involves the balancing of the needs of the physical environment with the needs of the community, and the needs of the tourist. However, this principle tends to imply that the needs of the tourist as a distinctive group are homogeneous, and the needs of the local community as a distinctive group are also relatively homogeneous. In reality, this is too simplistic and unlikely to be the case, as there are different types of tourists as described in Chapter 2, and communities can be divided on grounds of class, ethnicity, religion and gender.

Nevertheless, community control over development decision making has often been advocated by those who favour increased local democracy. Ecocentrics support development decision making at local level, not only on democratic principles, but also on the presumption that local people are more likely to act as stewards of the environment than external parties. However, community participation in planning and development may or may not be successful in encouraging people to favour less environmentally damaging development options, as attitudes to the physical environment are likely to reflect economic priorities. Even when tourism is presented to a local community as a less environmentally damaging development option than other forms of economic activity it may not be favoured by the local community. Burns and Holden (1995) comment upon the case of tourism development in the St Lucia Wetlands in Natal, South Africa, an area containing coral reefs, turtle beaches, high-afforested dunes, freshwater swamps, grasslands and estuaries. Rio Tinto Zinc (RTZ), the giant transnational mining corporation, wanted to mine the dunes for titanium dioxide slag. Despite assurances from RTZ over redressive environmental restoration of the area when the mining had ceased, central government was opposed to the use of the area for this purpose on environmental grounds, and instead favoured the development of ecotourism. However, local people, mainly Zulus, favoured the development of mining on the basis that RTZ had a good track record of paying comparatively high wages and investing in schools, clinics and other facilities. The Natal Parks Board, who run the surrounding game parks, were perceived by the local community as paying low wages, and having displaced local people from their lands to establish game reserves in the 1960s and 1970s.

Similarly, at Cairn Gorm, in the Scottish Highlands of Britain during the 1990s, there was a high level of controversy over the planned development of a funicular railway up the mountainside for the purposes of downhill skiing. Opposition to the scheme from major non-governmental organisations, such as the World Wildlife Fund (WWF) and the Royal Society for the Protection of Birds (RSPB), was based upon the possible environmental impacts on the arctic-alpine environment, which is unique within the British Isles. However, instead of receiving the support of the majority of local people, the two non-governmental organisations were largely seen as outsiders attempting to stop economic development to protect birdlife and flora, thereby denying local people employment and other economic opportunities. To realistically encourage stewardship of the environment by local communities, forms of tourism will need to be developed that are not only sympathetic to the environment, but also offer economic benefits as is the case in Senegal described in Box 6.5.

Box 6.5

Integrated rural tourism, Lower Casamance, Senegal

The development of tourism in the Lower Casamance is an example of how tourism can be used as a tool to enrich the livelihood and well-being of rural peoples. The aim of the scheme was to aid development, and also provide a more meaningful interchange between the local people and tourists than was being experienced on the coast, where hotels had largely been built with foreign capital and local people were excluded from tourist complexes by security guards and high walls.

In total, thirteen tourist camps have now been built, from an initial investment of US $7,000 each, provided by the l'Agence de Cooperation Culturelle et Technique. Tourists stay in simple lodges, built using traditional materials in the local architectural styles, so diminishing the differentiation between tourist and local facilities. Tourist numbers are restricted to a maximum of 20–40 guests and lodges are only constructed in villages where the population is 1,000 or above. Tourists eat locally grown produce wherever possible following traditional recipes. The scheme has proved to be a success, aiding development and social stability, improving health and educational facilities, and critically providing employment opportunities for the young which discourages them from migrating to larger towns to look for employment. Public expenditure of the revenues from tourism is controlled by village co-operatives.

Source: Gningue (1993)

In terms of a private sector led project aimed at encouraging sustainable tourism development, a notable project is the one developed by the International Federation of Tour Operators (IFTO), in the Balearic Isles, in the Mediterranean. Concerned with the problems that were being caused by tourism development in parts of Mallorca, the largest of the Balearic Isles, the ECOMOST ('Ecological Model of Sustainable

Tourism') study was supported by IFTO, the European Community, and the Ministry of Tourism for the Balearics. The aim of the ECOMOST project was to attempt to develop a 'model of sustainable tourism'. The researchers likened this model to tourism's stethoscope, the idea being that it could be used to gauge a destination's attainments against the ideal of sustainability, by using indicators of change. Within the context of the study the adopted definition of sustainable tourism was 'the maintenance of a balance where tourism runs at a profit but not at the expense of the natural, cultural, or ecological resources' (IFTO, 1994: 6). The report identifies four main requirements for the long-term maintenance of a tourism destination:

- the population should remain prosperous and keep its cultural identity;
- the place should remain attractive to tourists;
- nothing is done to damage the ecology; and
- an effective political framework.

The study developed 'check lists of dangers to sustainable tourism', under four different subheadings of 'population', 'tourism', 'ecology', and 'politics'. For each subsection, different components or targets were identified along with indicators of change, accompanied by 'critical values' which will determine if the indicator is positive or negative, as is illustrated in Box 6.6.

The idea of using indicators as an environmental management tool, to help achieve sustainable development, has gained pace since the adoption by the majority of the world governments of Agenda 21 in 1992. For example, the European Union is currently trying to establish relevant information systems, to ensure that socio-economic environments are developed in a way that respects the natural environment. Eurostat, the Statistical Office of the European Communities, in 1997 produced a list of forty-six indicators covering areas of the economy, society, environment and institutions. Examples of economic indicators that are relevant to the environment include the 'annual energy consumption per capita' and the level of 'environmental protection expenditure as a percentage of GDP'; examples of social indicators include the 'rate of growth of the urban population' and the 'per capita consumption of fossil fuel for motor vehicle transport'; examples of environmental indicators include the 'level of consumption of ozone depleting substances' and the 'quantity of fertiliser used in agriculture'; and an example of an institutional indicator is the level of 'expenditure on research and development as a percentage of GDP' (Eurostat, 1997). In 1998, the

Box 6.6

Example of possible indicators of sustainable tourism development

Topic	Component or target	Indicator	Critical value
Population	Preserving the population's prosperity	Population dynamics	If there is a continuous and major migration of the working population
		Unemployment rate	If greater than national average and/or increasing long term
		Per capita income	If lower than the national average
Tourism	Retaining satisfaction of guests and tour operators	Maintenance of quality and monitoring ecology	Persistent and/or significant criticism of destination's condition with special reference to: *quality* of accommodation, restaurants, service, leisure infrastructure; *overcrowding* of transport, beaches and sights; *ecological condition* of nature, the landscape and amounts of waste; *aesthetics* of townscape, landscape and cultural assets.
Ecology	Carrying capacity	Airport	If maximum capacity exceeded
		Tourist attractions	If roads and parking lots are continuously overcrowded at peak times
		Drinking water supplies	If there is: • water shortage in peak season • long term danger of salinity, floods, forest fires and other ecological damage
		Sewage	If European Union Standards for sewage disposal are mostly neglected
		Protection of species, use of protected areas	If, due to exploitation by tourists, flora and fauna is becoming imperilled or being destroyed

Box 6.6 continued

Topic	Component or target	Indicator	Critical value
		Pollution and emissions	If due to tourist use: • water • soil • air • health are continuously being threatened by pollution and/or noise
Politics	Effective tourism and ecologically oriented legislation	Not applicable	Existence of ecologically oriented quality standards

Source: Abridged and adapted from IFTO (1994: 10–15)

British government announced a range of indicators to help assess changes in the quality of life of the population, which could be used alongside traditional measures, such as gross domestic product (GDP), that are used to assess economic progress. Thirteen groups of indicators were decided upon covering the following areas: economic growth; social investment; employment; health; education and training; housing quality; climate change; air pollution; transport; water quality; wildlife; land use; and waste (Ward, 1998). However, although these developments are encouraging and indicators have undoubted use as part of an environmental management system, both for the general environment and tourism, their usefulness in securing a more sustainable future will ultimately be determined by the political views and wills of those in power.

Summary

● Sustainable development encapsulates a way of thinking about economic development that is inclusive of the environment. Development cannot take place upon a deteriorating environmental resource base, neither can the environment be protected when development excludes the cost of its destruction. The main aim of sustainable development is poverty alleviation, that is satisfying the needs of the world's population but achieving this in a way that does not threaten the earth's resources, nor the ability of future generations to satisfy their own needs.

● It is an ambiguous term and subsequently can be interpreted in different ways by groups with opposing political and philosophical viewpoints. Some groups and ideologies believe sustainable development can only be achieved by a radical restructuring of society involving changes in political structures and value systems. Others believe sustainable development can be achieved through alterations of the market system and improvement in technology that does not threaten the *status quo* in society.

● The application of the concept of sustainable development to tourism has also led to different perspectives on its meaning. A major difference is between those who interpret 'sustainable tourism' as advocating the sustaining of tourism in a destination, and those who advocate tourism as a vehicle for achieving sustainable development, which encompasses much wider socially determined goals and priorities. To date, the principles for sustainable tourism that have emerged from the tourism sector emphasise the former interpretation rather than the latter.

Further reading

Baker, S., Kousis, M., Richardson, D. and Young, S. (eds) (1997) *The Politics of Sustainable Development: Theory, Policy and Practice within the European Union*, London: Routledge.

Doyle, T. and McEachern, D. (1998) *Environment and Politics*, London: Routledge.

Mowforth, M. and Munt, I. (1998) *Tourism and Sustainability: new tourism in the Third World*, London: Routledge.

Pepper, D. (1993) *Eco-socialism: from deep ecology to social justice*, London: Routledge.

Reid, D. (1995) *Sustainable Development: An Introductory Guide*, London: Earthscan.

7 The future of tourism's relationship with the environment

- ⊙ **The growth of green consumerism and its effect upon tourism**
- ⊙ **The meanings of alternative tourism and ecotourism**
- ⊙ **The future of tourism's relationship with the environment**

Introduction

The raised awareness of the impacts society was having upon the global environment, including issues such as resource usage, pollution, animal and human rights, meant that by the end of the twentieth century some individuals were beginning to change their attitudes to consumerism and demand products that were more 'environmentally friendly' and 'ethically correct'. This change in buyer behaviour was given the name 'green consumerism'. This chapter considers the extent of the influence of green consumerism upon tourism, examines 'new' forms of tourism, and proceeds to consider the future of tourism into the twenty-first century.

The growth of 'green' consumerism

The late 1980s in the developed countries of the world marked the beginnings of a change in consumer behaviour, with concerns over the environmental effects of products being expressed in the buying patterns of a significant market segment of consumers. Cairncross (1991) takes the view that the extensive media coverage of environmental concerns, such as climatic change, the burning of rainforests and the vanishing ozone layer, was highly influential in getting people to think about the environmental effects of what they were buying. Martin (1997) takes the view that the rapid economic growth experienced in Britain during the early 1980s was influential in moving people's concerns on from basic economic needs to wider ethical considerations of consumerism. Generally, levels of public concern over the environment tend to reflect economic cycles, that is, when the economy is buoyant environmental concern is at its highest (Martin, 1997). Evidence of this shift in concern of consumerism towards environmental issues was demonstrated by the *Green Consumer Guide* becoming top of the list of best-selling books in Britain in September 1988. This publication was a seminal consumer guide, in that for the first time, ratings of the performance of companies and products were provided based upon environmental criteria.

Concern over the environment manifested itself in consumer behaviour, in various ways, in different countries. In Britain, there was a virtual consumer boycott of aerosols that contained chloro-fluorocarbons (CFCs), after the media and environmental pressure groups alerted consumers to the role of CFCs in ozone depletion (Swarbrooke and Horner, 1999). In America, consumers' primary concern was with waste and excessive packaging on products, and German consumers' concerns also rested with packaging and the plastics used to make drink bottles (Cairncross, 1991).

Market research at the end of the 1980s supported the observation that concern over the environment was high. Martin (1997) comments on the results of a survey conducted in 1989, of a representation of the British public by the Market Opinion Research Institute (MORI), a leading British market research company. When asked the question: 'What would you say is the most important issue facing Britain today?', 35 per cent of the respondents identified environmental issues, a higher percentage than for health care, unemployment, and inflation. However, as Martin points out, just sixteen months later when the survey was repeated, only 10 per cent of the sample identified the environment as being the most important issue and in subsequent surveys this would seem to be a consistent rating.

This 'outbreak' of green consumerism forced companies to begin to respond to the environmental concerns of an affluent and sizeable minority of the market. Not only did this influence the retail side of the market but also the supply side of the market as a few retailers began to check the environmental practices of their suppliers. For instance, B&Q a British hardware chain, questioned their suppliers about the quality of their environmental management of the peat bogs from which the peat sold as fertiliser in their shops came from, and one supplier was subsequently discontinued for having insufficient environmental standards. In America, the three leading tuna canners, controlling 70 per cent of the market, decided to ensure that their fish were caught only in 'friendly ways' which were not harmful to other types of fish (Cairncross, 1991).

Other new companies emerged into the consumer market, based upon a philosophy of selling ethically sound and environmentally friendly products. One of the most notable business success stories in the British market was the Body Shop, which now has hundreds of stores worldwide. The Body Shop's primary concerns are with skin and hair care, and it has emphasised from its foundation that environmental and social considerations have been at the forefront of its decision making. The view of the Body Shop is made clear in their policy document: 'The Body Shop has always had a clear, top-level commitment to environmental and social excellence. Because of this, we make sure we include environmental issues in every area of our operations' (Body Shop, 1992: 5). As part of this policy, the company does not carry out any animal experiments to test its cosmetics and also emphasises 'fair trade' practices with its suppliers, based upon buying from businesses who wish to protect their culture and practise traditional and sustainable land use. There was also a growing consumer demand for ethical investment funds, which can be traced to the early 1980s and the desire of the universities and churches in America not to put money into companies trading with South Africa, whose government supported the system of apartheid at that time (Cairncross, 1991).

The consequences for companies that ignore consumers' environmental concerns are illustrated by the case of Royal Dutch Shell, the world's biggest oil company. In 1995, Shell intended to bury an obsolete oil rig called the Brent Spar, at the bottom of the North Sea. Concerns raised by Greenpeace, the non-governmental organisation, over the possibilities that toxic sludge released from the rig would contaminate the sea floor, led to a great deal of controversy over the plan. Activists encouraged boycotts of Shell service stations in Germany and motorists also began to shun Shell garages in Denmark and Holland. In Germany, service station income fell 30 per cent (Hertsgaard, 1999) and the financial losses and adverse publicity led Shell to cancel the sinking of the oil rig.

At the beginning of the twenty-first century, 'green consumerism' seems to have established itself as an integral part of the consumer market. Although it is difficult to define exactly what green consumerism is, and therefore measure it, consumer surveys of behaviour patterns, based upon environmental attitudes, lend support to the feeling that environmental concern will play an increasing part in consumer behaviour in the future. For instance, when a sample of 2,000 people in Britain were asked if they had 'Selected one product over another because of environmentally friendly packaging/ formulation/ advertising', 36 per cent said they had in 1996 compared to 19 per cent in 1988 (Martin, 1997). Market research results in America also support evidence of a trend towards green consumerism. A survey by Michael Peters (MPG) of Americans in the summer of 1989, found that 53 per cent of the public had decided not to buy a product during the previous year, owing to concerns over the effects of the product or the packaging upon the environment. A survey carried out by the same firm in Canada found that approximately 66 per cent of the Canadian public would be more likely to buy a product that is recyclable or biodegradable, and approximately 60 per cent said they would be willing to pay more for such products (Cairncross, 1991). The growth in demand for organic farm products since the mid-1990s in Britain, associated with scares over salmonella poisoning in chickens and bovine spongiform encephalopathy (BSE) in cattle, also support the notion that people want to consume in a healthier, less environmentally damaging and more ethical way. The relatively high prices charged for organic farm produce in comparison to other produce reflect the willingness of a sizeable part of the consumer market, to pay premium prices for 'environmentally sound' produce.

The willingness of people to pay more for products and services that are advertised as being environmentally sensitive does, however, offer the opportunity for consumer exploitation by companies. Claims that a company operates in an 'environmentally friendly way' are often difficult to substantiate, and many people remain sceptical that a company's claim of employing green practices and acting in an environmentally conscious way, is little more than a public relations exercise. According to Cairncross (1991), a market survey carried out by Environmental Research Associates of Princeton, New Jersey, found that 47 per cent of consumers thought that environmental claims made by many companies were 'mere gimmickry'.

Consumers may often be confused about which products are genuinely environmentally friendly, given the lack of, or misleading, information that is sometimes presented. There is a necessity for the regulation of environmental claims made by companies through independent bodies. For example, organisations such as the Organic Soil Association in

Britain who verify farmer's claims of organic farming have been influential in implementing standards set by the European Commission on organic farming and reassuring the public that claims are genuine. Thus, although certification by the association is voluntary, it has become virtually necessary for any UK farmer wishing to sell organic produce to have certification, especially to sell their produce in supermarket chains.

In the tourism sector, besides the 'International Standards Organisation' (ISO) logo that is available to companies who implement 'environmental management systems', other green labels emphasising green credibility have appeared over recent years. One such scheme is the 'Green Globe', which was originally launched by the World Travel and Tourism Council (WTTC) in 1994 (Hawkins, 1997). The WTTC is a trade association which incorporates many of the leading multinational travel and tourism corporations. In 1999, 'Green Globe 21 Certification' became established as an organisation independent of the WTTC, concerned with verifying the environmental standards of companies and destinations operating in the tourism market.

The Green Globe 21 logo is applicable to many of the different elements of the tourism system, including tour operators, transport services, visitor attractions, hotels, local communities and destinations. Any travel and tourism company can apply to become a Green Globe 21 member as long as they want to be involved in sustainable tourism development. Green Globe 21 provide information and advice on the ways companies can modify their business practices to improve the environmental quality of their operations. After six months the company must decide whether to pursue formal certification from Green Globe 21, which involves establishing environmental management systems (EMS), necessitating the formulation of an environmental policy for the company, and environmental auditing of their business practices. Once the company decides to pursue certification they must pay a fee to Green Globe 21, which in 2000 varied from US $750 for local companies to US $15,000 for companies operating internationally, after which they can use the Green Globe 21 logo.

The verification that standards are being met is undertaken by Société Générale de Surveillance (SGS), an international verification body, and an International Advisory Council of industry, government, and NGO experts. This is a direct attempt to overcome the earlier criticism of the scheme, which was that as environmental monitoring was carried out on a self-assessment basis by the company rather than by independent verifiers, there was no real guarantee of standards and it was open to bias. However, a major criticism of the present scheme is that a company can use the logo after electing to follow the certification scheme, without having actually put an environmental management system into place. Once the company has implemented an EMS system and met the necessary criteria, a new logo is given to the company with a tick in it, though the ability of the consumer to understand the difference is very uncertain. Indeed, at present the level of awareness of the logo in the consumer market is probably low, and its meaning uncertain. Consequently, the scheme remains open to criticism about its standards, and to suspicion over the degree to which it represents a true desire to improve the environmental standards of the tourism industry, or an attempt to make money on the premise of voluntary environmental regulation.

Consumers are also encouraged to become members of Green Globe 21 by registering on their web page, for which there is no charge, but customers will find environmental information about companies and destinations. The third part of the system is the inclusion of communities into the award scheme. The basic cost of membership for communities is US $50,000 and thereafter the fee will be adjusted to reflect the complexity and scale of the EMS.

The scheme is innovative, both in the idea of certifying the operations of the tourism sector, and in its holistic approach which recognises tourism as a system, incorporating consumers, the industry, and the destination community. However, the scheme is very market-driven and its success will be dependent upon a growing number of consumers being interested in the green credentials of tourism companies and destinations, and having faith in the environmental integrity of the logo.

Another symbol of environmental standards in the tourism sector is the kitemark awarded by the 'Campaign for Environmentally Responsible Tourism' (CERT). Established in 1994, CERT is currently located in the United Kingdom and unfortunately is currently only accessible to UK tour-operating companies, although there are plans to expand the scheme into other countries and other companies in the sector. Tour operators can be awarded the CERT logo if they make a commitment to:

- develop and publish an environmental policy statement which is relevant to the business needs and destinations of operation;
- ensure that staff are aware of this environmental policy and trained in its implementation;

- distribute a lively, upbeat leaflet to clients which provides advice about how they can reduce the environmental impact of their holiday;
- include some questions about their environmental performance in their customer evaluation form. These are passed on to CERT and provide independent verification of the status and effectiveness of their environmental programme;
- help raise funds for environmental projects which will bring benefits to the environment, local people and the travel industry.

(Hawkins, 1997: 15)

The effectiveness of the company in meeting the above criteria is verified by a CERT specialist who also gives specialist advice to help the company. However, as for Green Globe 21, the success of this scheme will be dependent upon its recognition by consumers, and an understanding of and belief in the environmental quality that is portrayed in the logo.

Consumer trends and green tourism

The influence of green consumerism upon the tourism market is not easy to discern. As Swarbrooke and Horner (1999: 198) comment: 'One thing is clear – as the debate has developed, the term "green tourist" has not achieved the acceptance that the phrase "green consumer" has in general.' This does not mean that consumers do not have concerns about the activities of the tourism industry. The interests of consumers in the environmental aspects of tourism were reflected as early as 1991 with the publishing of the book, *The Good Tourist*, which advised tourists how to act in a more environmentally responsible way. As Wood and House (1991) comment on the back cover:

Thankfully now, the mistakes of the past are being recognised and a new breed of tourist is emerging. Travelling independently or in small groups, living as locals caring for the country they are visiting and its peoples, the Good Tourists take a long-term look at travelling and the world we live in.

Empirical research also supports the view that consumers are aware of the environmental impacts of the tourism industry. Martin (1997) cites a survey carried out by MORI in the summer of 1995 in Britain, in which when asked 'how much damage to the environment do you think travel and tourism causes?', 64 per cent of the respondents were found to think that tourism caused some degree of damage to the environment. However, in the same survey, when asked about the damage to the environment caused by tourism in comparison to other industries, tourism ranked fourteenth out of seventeen possible choices behind other services such as road transport and domestic waste disposal.

The opinion of the public about the efforts of the tourism industry to mitigate any of the harmful environmental impacts caused by it, in comparison to the efforts of other industries to reduce the negative environmental impacts their operations may be causing, is not reassuring for the sector. When presented with the statement, 'I'd like you to tell me how much you feel each industry is doing to reduce any harmful effects its activities might have on the environment', tourism ranked sixteenth out of seventeen industries behind 'nuclear fuel manufacturing and reprocessing' and 'oil exploration and production' (Martin, 1997). Yet, as for other consumer goods and services, there does seem to be a market opportunity for tourism companies to increase their revenues through environmental improvements. In the 1995 MORI survey, 30 per cent of the respondents stated they would be prepared to pay an extra premium on top of their holiday price, to ensure that tourism companies were committed to environmental protection. In reality, there is always a difference between people's willingness to say they would pay more for environmental care, and the likelihood of them actually doing it, as was discussed within the context of charging admission fees to visit national parks in Box 4.6. Nevertheless, the results suggest that a sizeable minority of the market is willing to pay more for environmental quality.

Although concerns over the environmental effects of tourism are unlikely to stop the vast majority of people taking some kind of holiday, for a part of the market, the type of holiday they choose may be increasingly influenced by their environmental attitudes. The development of green consumerism in the 1980s coincided with the development of a range of holiday types that inferred a greater level of awareness of the environment than is associated with mass tourism. Many terms have subsequently been used, often interchangeably, to describe these new forms of tourism including 'alternative', 'green', 'nature', 'sustainable', 'responsible', and 'ecotourism'. Mowforth and Munt (1998) extend this list to include other forms of new tourism such as academic tourism, agro-tourism, appropriate tourism, contact tourism, and wildlife tourism. A common theme of these labels is that they are indicative of a more caring approach to how tourism should be developed in the future. As Shackley (1996: 12) comments on new approaches to tourism: 'Terms such as environmentally friendly tourism, sustainable tourism, ecotourism, responsible tourism, low impact tourism are just a few among many in common use. . . . These designate low impact tourism programmes which might result in some form of sustainable benefits to the destination area.'

Certainly, the idea of alternative types of tourism to mass tourism seems to have found favour with a significant segment of the tourist market. Besides the growth in environmental consciousness since the late 1980s, the development of alternative forms of tourism can also be associated with consumer overfamiliarity with mass tourism, and a subsequent desire for new types of holidays. This latter point means that the concept of 'alternative tourism' can be interpreted in at least two ways: alternative tourism as a form of more environmentally aware tourism; or alternative tourism as types of tourism that are different to mainstream tourism without necessarily being any less environmentally damaging. For example, 'activity holidays' have become a segment of the tourism market that is growing, attracts high income groups, and is a component of a healthy lifestyle for many. Concerning changes in the market and the growth of the activity holiday market in Scotland, Scottish Enterprise (1995: 2) comment:

> Changes in consumer behaviour and values are a driving force for tourism. Today's tourist is becoming more experienced, more quality conscious, more independent, more discerning, and is therefore more difficult to please. . . . However, not only is the tourist becoming more sophisticated, experienced and quality conscious but also more active on holiday, reflecting the general growing awareness of healthy lifestyles and the increasing importance of active relaxation.

Yet it would be mistaken to think that a sports-based activity holiday would necessarily be compatible with the surrounding environment. For example, mountain biking, downhill skiing, and even hiking have all caused soil erosion, disturbance of wildlife, and problems of aesthetic pollution in mountain areas.

So what are the criteria of alternative tourism? Whilst there is no universal agreement on a definition of what alternative tourism actually is (Brown, 1998), the differences of alternative tourism to mass tourism are highlighted by Cater (1993: 85) as follows: 'Activities are likely to be small scale, locally owned with consequentially low impact, leakages and a high proportion of profits retained locally. These contrast with large-scale multinational concerns typified by high leakages which characterise mass tourism.' Utilising this definition it is possible to highlight the characteristics of alternative tourism which differentiate it from mainstream tourism as shown in Box 7.1.

Box 7.1

Characteristics of alternative tourism

- small scale of development with high rates of local ownership
- minimised negative environmental and social impacts
- maximised linkages to other sectors of the local economy, such as agriculture, reducing a reliance upon imports
- retention of the majority of the economic expenditure from tourism by local people
- localised power sharing and involvement of people in the decision-making process
- pace of development directed and controlled by local people rather than external influences

---

Using these criteria, alternative tourism surpasses purely a concern for the physical environment which typifies green tourism, to include economic, social, and cultural considerations. If the physical, environmental and cultural dimensions of the environment are considered in an integrated fashion, and tourism is developed with the characteristics displayed in Box 7.1, then alternative tourism can be viewed as being synonymous with the concept of sustainable tourism development.

A kind of tourism that is often associated with the characteristics of alternative tourism outlined above is 'ecotourism'. According to Shackley (1996) it is a term invented by conservationists in the 1970s, whilst according to Fennell (1999) the term can be traced back as far as 1965 to the work of Hetzer, who used it to explain the interaction between tourists and the environments that they come into contact with. However, like the concept of alternative tourism, there is no consensus about the meaning of ecotourism (Goodwin, 1996). The difficulty of trying to define ecotourism is referred to by Cater (1994: 3) thus: 'In particular, the term ecotourism is surrounded by confusion. Is it a form of 'alternative tourism' (furthermore, what is "alternative tourism?")? Is it responsible (defined in terms of environmental, socio-cultural, moral or practical terms)? Is it sustainable (however defined)?' Waldeback (1995) recognises different dimensions and interpretations of what ecotourism can be, as shown in Box 7.2.

Box 7.2

Dimensions of ecotourism

Activity – tourism which is based upon experiencing natural and cultural resources

Business – tour operators who provide ecotourism tours

Philosophy – a respect for land, nature, people and cultures

Strategy – a tool for conservation, economic development and cultural revival

Marketing device – for promoting tourism products with an environmental emphasis

Handle – convenient umbrella name for a number of tourism related concepts such as 'responsible or ethical travel', 'low-impact tourism', 'educational travel', 'green tourism' and so on

Symbol – of the debate about the relationship between tourism and the environment

Principles and goals – defining the symbiotic and sustainable relationship between tourism and the environment

Source: Adapted from Waldeback (1995)

Evidence of the confusion over the term can be seen in the continued attempts to define what it is. For example, McLaren (1998: 97) comments:

> Ecotravel involves activities in the great outdoors – nature tourism, adventure travel, birding, camping, skiing, whale watching, and archeological digs that take place in marine, mountain, island and desert ecosystems. Much of the travel is now called ecotourism, although critics argue that the definition of 'ecotourism' is so broad that almost any travel would qualify, as long as something green was seen along the way.

The danger of ecotourism being used as a marketing term in a disingenuous fashion to promote unsustainable forms of tourism is highlighted by Goodwin (1996), who comments upon its opportunist usage by the tourism industry, where the tag 'eco' has become synonymous with responsible consumerism. Hall and Kinnaird (1994) also make the point about the problematic nature of defining ecotourism, stating that there is no consistent definition. They comment that:

> this term [author's note: ecotourism] is used in a generic sense to
> cover tourism development which is sympathetic to, complements
> and/or is employed as a vehicle for, conserving and sustaining natural
> and cultural environments and their resources and which may
> encompass the domain of such terms as 'sustainable', 'green', 'soft',
> and 'alternative' tourism.
>
> (Hall and Kinnaird, 1994: 111)

Holden and Kealy (1996: 60) add an explicit economic component in
their definition:

> Implicit in all the definitions is respect or friendliness for the physical
> and cultural environment, i.e. developing a form of tourism that is
> non-damaging and non-degrading; subject to adequate and
> appropriate management controls; and that offers financial
> contributions for the protection of indigenous cultures and
> environments.

Goodwin (1996: 288) links this economic dimension directly to
conservation of resources in his definition of ecotourism as:

> Low impact nature tourism which contributes to the maintenance of
> species and habitats either directly through a contribution to
> conservation and/or indirectly by providing revenue to the local
> community sufficient for local people to value, and therefore protect,
> their wildlife heritage area as a source of income.

Both these latter definitions emphasise the economic aspect of
ecotourism in conservation, and the involvement, instead of the exclusion
of local communities in the conservation process. A comprehensive list
of guiding principles for the development of ecotourism is set out in Box
7.3 from Wight (1994: 40).

The direct relevance of ecotourism to sustainability following the
principles of Wight (1994) is highlighted by Shackley (1996: 13)
stating:

> Ecotourism projects should meet the following criteria. They must:
> - be sustainable (defined as meeting present needs without
>   compromising the ability to meet future needs)
> - give the visitor a unique and outstanding experience
> - maintain the quality of the environment.

Waldeback (1995) also suggests a range of goals that should be
accommodated within the process of ecotourism development:

Box 7.3

Guiding principles for ecotourism

- It should not degrade the resource and should be developed in an environmentally sound manner
- It should provide long-term benefits to the resource, to the local community and industry (benefits may be conservation, scientific, social, cultural, or economic)
- It should provide first-hand, participatory and enlightening experiences
- It should involve education amongst all parties – local communities, government, non-governmental organisations, industry and tourists (before, during and after the trip)
- It should encourage all-party recognition of the intrinsic values of the resource
- It should involve acceptance of the resource on its own terms, and in recognition of its limits, which involves supply-oriented management
- It should promote understanding and involve partnerships between many players, which could include government, non-governmental organisations, industry, scientists and locals (both before and during operations)
- It should promote moral and ethical responsibilities and behaviour towards the natural and cultural environment by all players

---

- sustainable use
- resource conservation
- cultural revival and decolonisation
- economic development and diversification
- life enhancement and personal growth
- maximum benefits and minimal costs/impacts
- learning about the natural culture and environment.

Who are the 'ecotourists'?

Based upon the principles and goals of ecotourism described by Wight (1994), Waldeback (1995) and Shackley (1996), it is evident that ecotourism places a much heavier emphasis upon conservation, education and ethics than mass tourism. The emergence of a form of tourism that is based upon the premise of an ethical relationship with the environment is reflective of changes that are occurring in the consumer market for tourism. Poon's (1993) concept of the 'new tourist' was referred to in

Chapter 2. According to Poon 'new tourists' display a 'see and enjoy but do not destroy' attitude and do not assert that the 'west is best'. Other labels have also been given to the tourists who constitute this emerging environmentally conscious market segment, including the 'ethical tourist', 'environmentally responsible tourist', 'good tourist' and 'ecotourist' (Swarbrooke and Horner, 1999).

So who are the ecotourists? Just as there is no definitive definition of ecotourism, similarly there are different opinions on the characteristics of ecotourists, and who ecotourists actually are. Given the difficulty in defining ecotourism, it is unsurprising that it is subsequently difficult to categorise the 'ecotourist', which leads to confusion over the type of behaviour that ecotourists could be expected to display. As Cater (1994: 76) comments: 'There is an inherent risk in assuming that the ecotourist is automatically an environmentally sensitive breed. Although small, specialist, guided groups of ecotourists may attempt to conform to this identity, the net has now been cast sufficiently wide to include less responsible behaviour.'

It would therefore be mistaken to think that being labelled an 'ecotourist' will necessarily mean that a tourist will particularly desire to be environmentally educated and have a minimal impact upon the environment. According to Mackay (1994), the tourist industry divides ecotourists into three distinct categories, of the big 'E', little 'E' and soft adventure types. The most popular group is the little 'E', in which tourists' environmental concerns are characterised by a wish to know that the hotel, airline or tour operator they intend to use has acceptable environmental standards. Big 'E' travellers are willing to travel into new, 'undiscovered' areas, and accept the standards of accommodation and services offered by local people or camp in the wilderness. The 'soft adventurer' also wishes to visit wilderness areas, but wishes to visit them in comfort, however, without a sense of feeling that the nature or culture of the area they are visiting is being 'exploited' through tourism.

A further categorisation of tourists into different types relating to their level of interest in the environment is shown in Figure 7.1, based upon the work of Cleverdon (1999).The level of demand for each typology is reflected in the width of the base of each segment, that is, demand decreases upwards from the base of the pyramid to its apex. The model suggests that the largest segment of the tourist market, the 'loungers', has a low level of interest in the environment beyond its providing pleasant surroundings. The focus of the holiday of this typology is likely to be

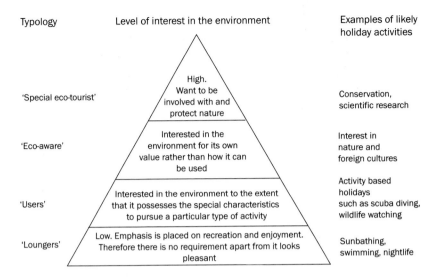

Typology | Level of interest in the environment | Examples of likely holiday activities

'Special eco-tourist' — High. Want to be involved with and protect nature — Conservation, scientific research

'Eco-aware' — Interested in the environment for its own value rather than how it can be used — Interest in nature and foreign cultures

'Users' — Interested in the environment to the extent that it possesses the special characteristics to pursue a particular type of activity — Activity based holidays such as scuba diving, wildlife watching

'Loungers' — Low. Emphasis is placed on recreation and enjoyment. Therefore there is no requirement apart from it looks pleasant — Sunbathing, swimming, nightlife

Figure 7.1 *Types of tourist, based upon their level of interest in the environment*
Source: After Cleverdon (1999)

based upon relaxing and enjoyment. The second typology of tourist, 'the users', are interested in the environment having the special features that are required for the type of holiday they wish to pursue. The types of environment required for this typology are therefore specialised and limited. For example, typical activities associated with this group would include wildlife watching and downhill skiing, each of which need environments that possess specialised features. The next typology, the 'eco-aware' show an increasing interest in the environment, not for how they can use it but for its own sake. They have an interest in environmental issues connected with tourism, and would look for evidence of environmental commitment, and perhaps certification of a high level of environmental practice from the tourism companies and suppliers they use. Typical types of holiday activity will reflect an interest in knowing more about the nature and the culture of the destination they are visiting. The last group, the 'specialist ecotourist' has a high level of commitment to the environment, to the extent that they want to actively protect it. This is reflected in the kinds of holiday they participate in, such as conservation holidays or scientific research.

Just as it as incorrect to talk about tourists as being one homogeneous group, Swarbrooke and Horner (1999) suggest that it is not possible to talk about the 'green tourist' as if they were one homogeneous group, preferring to refer to 'shades of green'. They suggest that the environmental commitment of tourists will be influenced by an amalgam of different factors, including: their awareness and knowledge of the issues associated with tourism and the environment; attitudes towards the environment in general; and the degree of fulfilment of other commitments in their life such as employment, housing and family needs. The different categories of green tourist and their associated environmental attitudes and actions are shown in Figure 7.2.

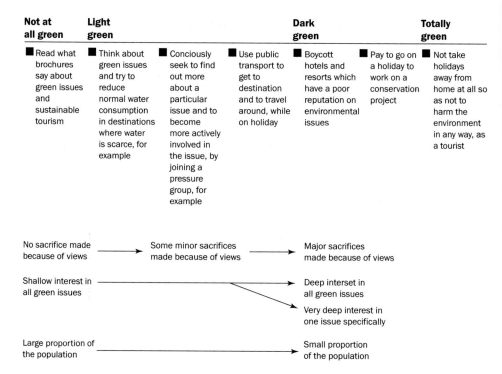

Figure 7.2 *Shades of green in tourism*

Source: J. Swarbrooke and S. Horner (1999) *Consumer Behaviour in Tourism,* Oxford: Butterworth-Heinemann, p. 202.

Empirical studies of ecotourists are limited, however research based upon demographic (age, gender, occupation, income) characteristics and psychographic (values, attitudes and motivation) characteristics has been attempted. According to Wearing and Neil (1999) research conducted in the USA suggests that ecotourists have higher than average incomes and levels of education, and are also willing to spend more than the normal tourists. In terms of their psychographic characteristics, Wearing and Neil report that they possess an environmental ethic, and are biocentric rather than anthropocentric in orientation.

'Ecotourism: product or principle?': the dangers of ecotourism and alternative tourism

The title 'ecotourism: product or principle?' is a question that was raised by Cater and Lowman (1994) to illustrate the different interpretations of the term 'ecotourism'. Although the preceding part of this chapter has illustrated how ecotourism may be defined using a set of principles which reflect very closely those of the concept of sustainable tourism, as previously stated, there is also a distinct danger that ecotourism may be interpreted purely as another tourism 'product' to be sold into the market-place. This may involve putting landscapes, that have had little or no human interference with them, directly into the market-place for tourism. As Baroness Chalker, then Minister for Overseas Development for the United Kingdom government commented, one of the characteristics of ecotourism is 'attracting tourists to natural environments which are unique and accessible' (1994: 90). Given that many of the most beautiful natural environments are located in rural areas of less developed countries, where often there are problems of underdevelopment and poverty, the economic incentives to be gained from ecotourism mean that there is a possible threat to the conservation of these areas. This is a theme developed by Butler (1990) who questions, profoundly, the idea that forms of 'alternative' tourism offer an alternative approach to mass tourism. As he points out, the idea of 'alternative tourism' sounds good to everyone and is hard to disagree with, but is so vague in its meaning that like sustainable development, it can mean almost anything to anyone. Further, he continues to say that although mass tourism maybe criticised a lot, many people still like to be mass tourists. It is therefore highly improbable that low-scale alternative tourism will ever replace the mass tourism market. Butler is especially critical of the idea that one form of tourism can be promoted as the panacea to problems that have resulted from extensive and long-term tourism development, commenting: 'Mass tourism need not be uncontrolled, unplanned, short-term or unstable' (Butler, 1990: 40).

Harrison (1996) also questions the extent to which ecotourism and 'good tourists' are genuinely alternatives to mass tourism or whether they merely represent stages towards it. The danger of being complacent over ecotourism is also highlighted by De Alwis (1998: 232), who cites Robertson Collins the tourism veteran and conservationist as saying:

> Most of those who tag tourism 'ecotourism', after a while begin to believe that all they do is ecologically and people friendly. The customer who buys into an ecotourism labelled product also satisfies his or her conscience that they are not causing any damage to the nature or people that they visit. . . . Ecotourism today, is big business. Products with an eco label are able to command higher prices in the market place. Similarly, there is also substantial funding for 'ecotourism' proponents by donors and well meaning agencies, making it a lucrative consultancy based business as well. This situation has unfortunately led to the creation of the popular perception that 'tourism is bad' and 'ecotourism is good'.

De Alwis (1998) also makes the point that demand for ecotourism is market driven, as more people express an interest in nature, and the desire for independent travel has become stronger. The dangers of ecotourism are also elaborated by Shackley (1996) who suggests that the introduction of visitors into ecosystems that we barely understand can instigate long-term changes. Rogers and Aitchison (1998) draw attention to the point that, although the definitions of ecotourism sound attractive in theory, they do not address the more problematic issue of how to develop and promote ecotourism. They point out that ecotourism is often subject to the same market forces that led to the environmental problems associated with mass tourism, and that there are many different players involved in both the public and private sectors, leading to problems of co-ordination and planning of development. They add that ecotourism sites are diverse and at early stages of development, which limits the opportunities for comparative study, and that models of 'good practice' are lacking.

The ability to make financial profit from ecotourism is also emphasised by Mowforth and Munt (1998). In their view not only does 'ecotourism' signal an interest in the 'environment' (*eco*logy) but it also indicates the ability to pay the high prices that such holidays command (*eco*nomic capital) . Ecotourism is, as has been explained, a form of tourism that relies upon being alternative to the mainstream, and is therefore in one sense exclusive. The association of ecotourism with exclusivity is a theme emphasised by Wheeller in his extensively quoted work (1993a, 1993b), in which he renames ecotourism as 'egotourism'. Mowforth and Munt (1998) link the idea of 'egotourism' to being in a competition for distinction, uniqueness and differentiation, alternatively expressed as another form of 'conspicuous consumption'. Given the seeming willingness of ecotourists to pay premium rates to visit 'unspoilt' nature, there are therefore strong economic and financial incentives for governments and business entrepreneurs to facilitate the use of the environment for ecotourism.

One of the great dangers in opening up new areas of the environment to ecotourism relates to the issue that unless development is strictly controlled, there is little reason to assume the development cycle should be any different than for other forms of tourism. There has been and perhaps still is an inability to accept that ecotourists are a form of 'trail-blazer', opening up new areas to the hordes they disdain, unless there are strict development and planning controls (Burns and Holden, 1995). The application of planning controls limiting the number of ecotourists to a particular area may in reality prove difficult to implement and maintain, and may fluctuate in response to a country's economic situation and the policy of the government in power. This latter point is illustrated in Box 7.4 concerning the development of ecotourism in Belize.

Box 7.4

Ecotourism in Belize

One of the first countries to promote itself as an ecotourism destination was Belize in Central America, which is less than two hours' flying time from Miami in the USA. The country has a small land area of 23,000 square kilometres but contains a wide variety of vegetation types, including mangrove swamps, wetland savanna, mountain pine forests, and tropical rainforests. Additionally, the coral reef off Belize's coast is the second longest in the world, after the Australian Barrier Reef. There are also several notable archaeological sites of the Mayan civilisation. The attraction of ecotourism for the Belize government is to earn foreign exchange, and the government committed itself to balanced and environmentally sound tourism in the 1980s. The typical advertising slogans to promote Belize are 'Belize so natural', 'friendly and unspoilt', 'naturally yours'. The majority of tourists who come to Belize do not arrive on package tours, but their travel and accommodation is organised by foreign tour operators, and subsequently the economic leakage factor is likely to be high. The financial opportunities to be gained from ecotourism have led to rapid property inflation around the coastline and it is now estimated that 90 per cent of the coast line is under foreign ownership. United States based developers are building luxury resorts, with golf courses and marinas, and it is estimated that 65 per cent of the members of the Belize Tourism Industry Association are expatriots mainly from the USA. However, the government has no plans to restrict foreign ownership of land as it requires the foreign investment. The arrival of tourists at

continued

approximately 200,000 per annum is also beginning to have environmental consequences. Some of the coral in the Hol Chan Marine Reserve established in 1987 has become infected with black band disease, a form of algae which kills the coral, when it has already been broken. There has also been a decline in the numbers of conch and lobster owing to overfishing, partly to satisfy tourism demands. Other examples of the impacts of tourism are more dramatic. In 1992, 'Eco-terrorism at Hatchet Cay' was the headline, in reference to a US resort owner who had attempted to blow up part of the coral reef to make his resort more accessible to visiting boats. Unfortunately, it would seem that ecotourism development in Belize has become characterised by the problems of mass tourism, including foreign exchange leakages, a high degree of foreign ownership of prime coastal land and tourism facilities, and environmental degradation. The government's desire for foreign exchange, combined with global market forces and trade liberalisation, means that the environment is becoming little more than a selling point for a form of mass ecotourism.

Sources: Cater (1992), Panos (1995) and Mowforth and Munt (1998)

Other criticisms have also been made of alternative forms of tourism like ecotourism, such as that they cannot provide a viable level of income and employment, and that education of the tourist into a form of behaviour that is respectful of the environment is a long-term process for which the financing and logistics have not been thought through properly (Brown, 1998).

It is evident that alternative forms of tourism will not necessarily lead to tourism that is culturally and environmentally friendly. In a neo-liberal global economy, the regulation of market forces to allow a balanced use of environmental resources will be critical in determining the success of any alternative tourism policy. However, this may not be that easy to achieve, especially for countries who are in need of foreign investment and foreign exchange. This means that for many less developed countries, the global political economy, which incorporates international debt repayment and the domination of transnational corporations, places their environments under threat of exploitation for short-term need and profits. The use of environmental resources for ecotourism may, in terms of negative environmental consequences, prove to be little different to that of mass tourism, in fact perhaps even more detrimental as ecotourism environments are likely to be rich in biodiversity and highly susceptible to damage.

The future relationship between tourism and the environment: to the year 2020

A central theme of this book has been that understanding the relationship that exists between tourism and the environment is concerned not just with the destinations that tourists go to but also the societies which they come from. It would seem that as societies become more economically developed, wealthier and urbanised, then the propensity to consume through tourism increases. The projected levels of tourism demand of 1,600 million international tourist arrivals by 2020 (World Tourism Organisation, 1998a) suggest that tourism will become an even more significant feature of society and the global economy than it is at present.

This projected growth in tourism will provide both opportunities and threats to the environments and societies it interacts with. The economic opportunities offered by tourism mean it is increasingly likely to feature as a key part of many governments' economic policies, especially in less developed countries. Importantly, by encouraging development and helping to eliminate poverty, tourism can assist in building a more sustainable future by meeting human needs and helping to conserve environments and wildlife, protecting them from threats such as poaching or potentially destructive forms of economic activity such as mining or logging. However, the achievement of this relies upon government policy reflecting a stronger commitment to an environmental ethic and a long-term perspective, than has historically been the case. The significant lesson of tourism development in the second half of the twentieth century was that although tourism can bring economic benefits, it can also contribute to the destruction of the natural environment. It can also be a cause of cultural changes, for instance changing the value systems of traditional societies by propagating consumerism and associated materialist values, and also displacing people from their traditional lands and denying them access to resources they require to meet their needs. Many of the negative effects of tourism have resulted from a *laissez-faire* approach to development, determined by free market forces, in which the full social costs of tourism development have failed to be reflected. Government policy and planning for tourism in the future is essential. Policy and planning need to reflect a balanced approach to how natural resources are used and include local communities in the development process. For want of a better word, a more 'sustainable' approach to tourism development is required.

As a key stakeholder and the catalyst to tourism development, the tourism industry will also need to increasingly address its role and responsibilities to the environment. The tourism industry continues to display increasing signs of market maturity, with the acquisition and merging of companies into larger multinational conglomerates, which are increasingly being floated on stock-exchanges as public shareholding companies. This trend is likely to continue as the market demand for tourism continues to increase. These multinational companies will subsequently have increasing influence in many destinations and markets around the world, and interact with the many locally owned small and medium-sized enterprises, which continue to constitute the major part of the supply market in most tourism destinations. In an increasingly global business environment and market-place, which facilitates the movement of international capital and the establishment of business operations in different countries, it can be argued that major international companies have an increasing responsibility to address ethical issues of their operations. Such issues will include their interaction with local communities, indigenous cultures, and the natural environment. Practical initiatives could include actions such as strengthening supply linkages to locally owned businesses, helping to reduce the level of economic leakages and increase the economic benefits of tourism for members of the local community, and developing environmental management systems to cover all aspects of their operations.

However, there is no one model that can act as blueprint of how tourism, the natural environment and local communities can interact in the most beneficial fashion in the future. Different destinations will have differing degrees of tolerance to tourism, based upon their own environmental characteristics, and the extent to which the natural and cultural environments have already been subjected to change by outside influences. Some tourism destinations, such as many of those on the western Mediterranean coastline, are already in a mature stage of development. For these areas, the major physical and cultural changes that tourism can bring have already been experienced, and in many of these areas efforts are being made to redress the negative effects of tourism. Other environments have yet to be 'discovered' by tourism, subsequently different environments will have different tolerance levels to tourism. This is not to say that destinations in a mature stage of development are free from any further environmental damage from tourism, as continued expansion may push them past a further threshold level, where environmental management and technological measures cannot redress the balance. However, it is particularly in areas of the world where tourism is developing, or has yet to develop, that there is the potential for many environmental problems. As the pressure grows for kinds of tourism that are increasingly based upon the natural environment, including special ecosystems, wildlife and indigenous cultures, there is a requirement for strict planning controls to avoid unacceptable levels of environmental change.

A further issue in defining tourism's future relationship with the environment is that ultimately what is an acceptable level or unacceptable level of environmental change and development will not be quantitatively fixed but determined by decision makers in society. Tourism development and economic growth cannot take place without some degree of trade-off between the use of the environment and economic benefits. Ultimately the extent of this trade-off, and the amount of environmental change incurred and whom it benefits, will be a reflection of decision makers' value systems and environmental philosophy. There is therefore a danger that where decisions are made by a small elite, there will be an inability to hear a wide range of views across the political and philosophical spectrums, and decisions will not be made on a consensual basis. Subsequently, there are likely to be clashes over the use of natural resources for tourism, particularly in situations where local people are denied access to resources that they have traditionally used to meet their needs. Tourism thus becomes viewed by many local people not as a constructive force for development but as a propagator of inequality. It is therefore necessary that the process of tourism development incorporates community participation, a principle agreed to by the majority of the world's governments when they became signatories of Agenda 21 in 1992.

Yet the future of tourism's relationship with the environment lies not only in what is happening in the tourism destination. Mass participation in international tourism is a feature of economically advanced countries, reflecting privilege, and is now an integral part of a postmodern consumer lifestyle. Whatever the motivations for participation in tourism, whether as consequence of anomie and the need for ego-enhancement or the desire for escapism *vis-à-vis* the search for the authentic, tourism is a function of economically advanced and urbanised societies. The desire for tourism is a reflection of the quality of life experience of many people living in western society. Empirical evidence would suggest that, whilst material wealth is higher than ever before in western countries, the quality of life in many of them has begun to decrease. For example

Carley and Spapens (1998) point out that according to the Index of Social Health published by Fordham University in New York which measures well-being, America's social health declined from 73.8 out of a possible 100 in 1970 to 40.6 in 1993 as homicides, suicides, the gap between the rich and the poor and drug use all increased. Likewise, the pressures of the consuming life are also evident in Britain. Under the title 'Britain in 2010: rich but far too stressed to enjoy it', the point is made that the pressure of modern working patterns and the desire for wealth are fuelling family breakdowns, drug and alcohol dependency. At the end of the twentieth century nearly a quarter of all men and women in Britain believed that they had sacrificed seeing their children grow up because of the pressures of work (Watson-Smyth, 1999).

Taking the view that tourism is representative of a wish to escape from societies, at least periodically, it can be interpreted as a symptom of postmodern society. It is likely that within the time-constrained framework of modern life, tourism will, for the majority of the population, be an experience to be consumed in a highly packaged format. Under these circumstances, the extent to which tourism can act as an educational medium, involving constructive and meaningful exchanges with foreign cultures and environments, is doubtful. Tourism's function for many in society will be to act as a reward for the stresses of day-to-day living, during the rest of the year.

As the world becomes increasingly urbanised and economically developed, and more people seek to consume the experiences and rewards of tourism, a major question facing tourism planners is where the predicted extra 1 billion tourists in the next twenty years are going to be accommodated. This question of course oversimplifies the issue, as within that figure of 1 billion lies a complex market pattern, including issues such as types of tourism, repeat visitations, and low and high seasons. Nevertheless, this rapid increase in the level of international tourism during the next twenty years will pose a strain on environmental and cultural resources. Many of these extra tourists will be travelling by air transport with the subsequent effects upon the atmosphere from the emissions of carbon and nitrogen oxides. Yet it is particularly in tourism destinations where the pressures on the environment will be most intense, as this is where tourism is most concentrated. Of particular concern is the increasing demand for nature tourism, which potentially could take tourists to some of the most ecologically fragile areas of the earth. As already mentioned, alternative tourism such as ecotourism may in reality represent little more than the first stage of mass tourism. If tourists can go to space, which will probably happen in the next five years, no part of the earth's environment can be considered to be inaccessible to tourism.

The relationship between tourism and the environment therefore faces many challenges over the next twenty years. Economic development and the social conditions of postmodern societies are a fertile breeding ground for tourism demand. The answer to a harmonious relationship in the future between tourism and the environment thus lies as much within the holistic philosophy of society and environmental attitudes as it does with a reductionist approach of finding technical solutions to environmental problems caused by tourism. The emergence of green consumerism since the late 1980s points to a realisation, at least by a part of the population, that present rates of consumption cannot continue without adverse environmental effects. At the beginning of the twenty-first century, we are aware that many adverse environmental effects are the consequence of human actions, such as ozone depletion, global warming, pollution of rivers and acid rain. There is increasingly a realisation, judging by the demand for more environmentally friendly products and ethical investment funds, that individual action can make a difference to environmental issues such as pollution and fair trade.

Some people are also increasingly disenchanted with the 'rat race' associated with 'comparative consumption', in which we work harder to buy more goods and services, to keep pace with our neighbours. The idea of 'downsizing' to a more simple lifestyle is an option a few people have decided to take in the western world, perhaps indicative of wider changes to come in the twenty-first century. However, for the majority in the next twenty years, tourism is probably going to be increasingly viewed as an essential component of having a better quality life. During the next twenty years tourism will bring changes to many environments around the world. In 2020, the extent to which these changes will be determined as positive or negative will be a reflection of the values held in society at that time and who is doing the judging.

Summary

- The emergence of green consumers in postmodern society represents a change in values of a significant part of the market towards consumerism. The growth in demand for items such as organic foods, cosmetics that are free of animal testing, fair trade items and ethical investment funds suggests that ethical and environmental concerns are becoming more important for some people in society. The extent to which this consumer concern has materialised in the tourism market is limited.

- There is a growing demand to visit environments that are regarded as being 'natural' and 'unspoilt'. Part of this demand is associated with the increasing levels of urbanisation and associated stress in society, which have cut people off from nature. One type of tourism that is increasingly talked about is 'ecotourism'. However, ecotourism is about much more than visiting nature, being based upon a set of principles advocating ethical approaches to tourism development and visitation.

- The increasing demand for nature based tourism is a source of concern. This concern is caused by the realisation that, without strict planning controls, there is little reason why what is commonly referred to as 'alternative tourism' should be anything more than the first step towards mass tourism. Many people also like to be mass tourists, and a key challenge for tourism planners in the future will be how to make mass tourism more sustainable.

Further reading

Now is the time to explore the bibliography!

Bibliography

Allaby, M. (ed.) (1994) *The Concise Oxford Dictionary of Ecology*, Oxford: Oxford University Press.

Badger, A., Barnett, P., Corbyn, L. and Keefe, J. (1996) *Trading Places: Tourism as Trade*, London: Tourism Concern.

Baker, S., Kousis, M., Richardson, D. and Young, S. (eds) (1997) *The Politics of Sustainable Development: Theory, Policy and Practice within the European Union*, London: Routledge.

Balmer, D. (2000) 'Take-off for a lesson in caring tourism', the *Observer* ('Escape' supplement), London, 23 January, p. 4.

Barke, M. and France, L.A. (1996) 'The Costa del Sol', in Barke, M., Towner, J. and Newton, M.T. (eds) *Tourism in Spain: Critical Issues*, Wallingford: CAB International, Ch. 11, pp. 265–308.

Barke, M. and Towner, J. (1996) 'Exploring the History of Leisure and Tourism in Spain', in Barke, M., Towner, J. and Newton, M.T. (eds) *Tourism in Spain: Critical Issues*, Wallingford: CAB International, Ch. 1, pp. 3–34.

Barker, L.M. (1982) 'Traditional Landscape and Mass Tourism in the Alps', *Geographical Review*, 72(4): 395–415.

Bartelmus, P. (1994) *Environment, growth and development: the concepts and stategies of sustainability*, London: Routledge.

BAT (1993) *European Tourism Analysis*, Hamburg: BAT-Leisure Research Institute.

Bayswater, M. (1991) 'Prospects for Mediterranean beach resorts: an Italian case study', *Tourism Management*, 5: 75–89.

Becheri, E. (1991) 'Rimini and Co – the end of a legend?: Dealing with the algae effect', *Tourism Management*, 12(3): 229–35.

Beeton, S. (1997) '*Visitors to National Parks: attitudes of walkers to commercial horseback tours*', Paper given at the Trebbi Conference, July 6–9, Sydney, Australia.

Bird, B.D.M. (1989) *Langkawi: from Mahsuri to Mahathir: tourism for whom?*, Selangor, Malaysia Institute of Social Analysis.

Bocock, R. (1993) *Consumption*, London: Routledge.

Bodlender, J.A. and Ward, T.J. (1987) *An Examination of Tourism Investment Incentives*, London: Horwath and Horwath.

Body Shop (1992) *The Green Book*, West Sussex: Body Shop International.

Boo, E. (1990) *Ecotourism: the Potentials and Pitfalls*, Vol. 1, Washington: World Wide Fund for Nature.

Boorstin, D.J. (1961) *The Image: A Guide to Pseudo-Events in America*, 1st edn, reprinted in (1992), New York: Vintage Books.

Booth, D.E. (1998) *The Environmental Consequences of Growth*, London: Routledge.

Bowcott, O., Traynor, I., Webster, P. and Walker, D. (1999) *Analysis: Green Politics*, the *Guardian*, March 11, p. 17.

Briguglio, L. and Briguglio, M. (1996) 'Sustainable Tourism in the Maltese Isles', in Briguglio, L., Butler, R., Harrison, D. and Filho, W.L. (eds) *Sustainable Tourism in Islands and Small States*, London: Pinter, Ch. 9, pp. 161–79.

Brown, D.O. (1998) 'Debt-funded Environmental Swaps in Africa: Vehicles for Tourism Development?', *Journal of Sustainable Tourism*, 6(1): 69–77.

Budowski, G. (1976) 'Tourism and Conservation: Conflict, Coexistence or Symbiosis', *Environmental Conservation*, 3: 27–31.

Bull, A. (1991) *The Economics of Travel and Tourism*, London: Pitman.

Burac, M. (1996) 'Tourism and Environment in Guadeloupe and Martinique', in Briguglio, L., Butler, R., Harrison, D. and Filho, W.L. (eds) *Sustainable Tourism in Islands and Small States: Case Studies*, London: Pinter, Ch. 4, pp. 63–74.

Burns, P. and Holden, A. (1995) *Tourism: A New Perspecive*, Hitchin: Prentice-Hall.

Burns, P. (1999) *An Introduction to Tourism and Anthroplogy*, London: Routledge.

Butler, R. (1990) 'Alternative Tourism: Pious Hope or Trojan Horse?', *Journal of Travel Research*, 28(3): 40–5.

Butler, R. (1993) 'Pre- and post impact assessment of tourism development', in Pearce, D.W. and Butler, R.W. (eds) *Tourism Research: Critiques and Challenges*, London: Routledge, Ch. 8, pp. 135–54.

Butler, R. (1997) 'The Concept of Carrying Capacity for Tourism Destinations: Dead or Merely Buried?', in Cooper, C. and Wanhill, S. (eds) '*Tourism Development: Environmental and Community Issues*', pp. 11–22.

Butler, R. (1998) 'Sustainable tourism – looking backwards in order to progress', in Hall, M.C. and Lew, A.A. (eds) *Sustainable Tourism: A Geographical Perspective*, Harlow: Longman, Ch. 3, pp. 25–34.

Cairncross, F. (1991) *Costing the Earth*, London: The Economist Books.

Carley, M. and Spapens, P. (1998) *Sharing The World: Sustainable Living and Global Equity in the 21st Century*, London: Earthscan Publications.

Cater, E. (1992) 'Profits from Paradise', *Geographical Magazine*, 64(3): 17–20.

Cater, E. (1993) 'Ecotourism in the Third World: Problems for Sustainable Tourism Development', *Tourism Management*, April, 85–90.

Cater, E. (1994) 'Ecotourism in the Third World – Problems and Prospects for Sustainability', in Cater, E. and Lowman, G. (eds) *Ecotourism: A Sustainable Option*, Chichester: Wiley, pp. 69–85.

Cater, E. and Lowman, G. (1994) *Ecotourism: a Sustainable Option*, Chichester: Wiley.

Clarke, J. and Critcher, C. (1985) *The Devil Makes Work: Leisure in Capitalist Britain*, Basingstoke: Macmillan.

Cleverdon, R. (1999) *Lecture Notes*, Centre for Leisure and Tourism Studies, University of North London.

Coccossis, H. and Parpairis, A. (1996) 'Tourism and Carrying Capacity in Coastal Areas: Mykonos, Greece', in Priestley, G.K., Edwards, J.A. and Coccossis, H. (eds) *Sustainable Tourism: European Experiences*, Wallingford: CAB International, Ch. 10, pp. 153–75.

Club Freestyle (1999) *Summer '99: Have It Your Way*, 2nd edn, London: Thomson Holidays.

Cohen, E. (1972) 'Towards a Sociology of International Tourism', *Social Research*, 39(1): 164–89.

Cohen, E. (1979) 'A Phenomenology of Tourist Experiences', *Sociology*, 13: 179–201.

Cohen, E. (1995) *Contemporary Tourism – Trends and Challenges: Sustainable Authenticity or Contrived Post-modernity?*, in Butler, R. and Pearce, D. (eds) *Change in Tourism: People, Places, Processes*, London: Routledge, Ch. 2, pp. 12–29.

Collin, P.H. (1995) *Dictionary of Ecology and Environment*, 3rd edn, Teddington: Peter Collin Publishing.

Coker, A. and Richards, C. (eds) (1992) *Valuing the Environment*, Chichester: Wiley.

Cooper, C., Fletcher, J., Gilbert, D., Wanhill, S. and Shepherd, R. (1998) *Tourism: Principles and Practices*, 2nd edn, Harlow: Longman.

Dalen, E. (1989) 'Research into values and consumer trends in Norway', *Tourism Management*, 10(3): 183–6, cited in Swarbrooke, J. and Horner, S. (1999) *Consumer Behaviour in Tourism*, Oxford: Butterworth-Heinemann.

Dann, G. (1977) 'Anomie, Ego-Enhancement and Tourism', *Annals of Tourism Research*, 4(4): 184–94.

Davidson, R. (1993) *Tourism*, 2nd edn, London: Pitman.

De Alwis, R. (1998) 'Globalisation of Ecotourism', in East, P., Luger, K. and Inmann, K. (eds) *Sustainability in Mountain Tourism: Perspectives for the Himalayan Countries*, Delhi: Book Faith India, pp. 231–6.

Department of the Environment (1991) *Tourism and the Environment: Maintaining the Balance*, London: HMSO.

Department for International Development (1997) *Tourism, Conservation and Sustainable Development: Comparitive Report*, Vol. 1, London: Department for International Development.

Douthwaite, R. (1992) *The Growth Illusion*, Bideford, Devon: Green Books.

Doyle, T. and McEachern, D. (1998) *Environment and Politics*, London: Routledge.

Drumm, A. (1995) *Converting from Nature Tourism to Ecotourism in the Ecuadorian Amazon*, Paper given at the World Conference on Sustainable Tourism, Lanzarote, April.

Eadington, W.R. and Smith, V.L. (1992) 'The Emergence of Alternative Forms of Tourism', in Smith, V. L. and Eadington, W.R. (eds) *Tourism Alternatives: Potentials and Problems in the Development of Tourism*, Philadelphia: University of Pennsylvania Press.

Edington, J. and Edington, A. (1986) *Ecology, Recreation and Tourism*, Cambridge: Cambridge University Press.

Elliott, J.A. (1994) *An Introduction to Sustainable Development: The Developing World*, London: Routledge.

English Tourist Board (1991) *Tourism and the Environment: Maintaining the Balance*, London: English Tourist Board.

European Tourism Analysis (1993) *Ten Main Characteristics for Quality Tourism*, Hamburg: BAT-Leisure Research Institute.

Eurostat (1997) *Indicators of Sustainable Development*, Luxembourg.

Evans, G. (1993) 'Tourists rush for kill a seal pup holiday', *Evening Standard*, London, July 5, p. 10.

Ezard, J. (1998) 'Ship Ahoy', the *Guardian*, London, March 23, p. 8.

Ezard, J. (1999) 'Ape rights today, and tomorrow a haddock may be my brother', the *Guardian*, London, February 2, p. 1.

Farrell, B. (1992) 'Tourism as an Element in Sustainable Development: Hana, Maui' in Smith, V. L. and Eadington, W.R. (eds) *Tourism Alternatives: Potentials and Problems in the Development of Tourism*, Philadelphia: University of Pennsylvania Press, Ch. 5, pp. 115–32.

Fennell, D.A. (1999) *Ecotourism: an Introduction*, London: Routledge.

Foster, J.B. (1994) *The Vulnerable Planet: A Short Economic History of the Environment*, New York: Cornerstone Books.

Friends of the Earth (1997) *Atmosphere and Transport Campaign*, www.foe.co.uk.

Gallup (1989) *American Express Global Travel Survey*, Princeton: Gallup Organisation Incorporated.

Gamero, E. (1992) 'Legislation for Sustainable Tourism: Balearic Islands', in Eber, S. (ed.) *Beyond the Green Horizon*, Godalming: World Wide Fund for Nature.

Gill, R. (1967) *Evaluation of Modern Economics*, New Jersey: Prentice Hall.

Gningue, A.M. (1993) 'Integrated Rural Tourism Lower Casamance', in Eber, S. (ed.) *Beyond the Green Horizon: a Discussion Paper on the Principles for Sustainable Tourism*, Godalming: World Wide Fund for Nature.

Goodall, B. (1994) 'Environmental Auditing: Current Best Practice' in Seaton, A.V., Jenkins, C.L., Wood, R.C., Deike, P.U.C., Bennett, M.M., Maclellan,

L.R. and Smith, R. (eds) *Tourism: The State of the Art*, Chichester: Wiley, Ch. 68, pp. 655–64.

Goodall, B. and Stabler, M.J. (1997) 'Principles Influencing the Determination of Environmental Standards for Sustainable Tourism' in Stabler, M.J. (ed.) *Tourism and Sustainability: Principles to Practice*, Wallingford: CAB International, Ch. 19, pp. 279–304.

Goodpaster, K.E. in Werhane, P.H. and Freeman, R.E. (1998) *Encyclopedic Dictionary of Business Ethics*, Oxford: Blackwell, pp. 51–7.

Goodwin, H. (1996) 'In Pursuit of Ecotourism', *Biodiversity and Conservation*, 5: 277–91.

Gosling, D. (1990) 'Religion and the Environment', in Angell, D.J.R., Comer, J.D. and Wilkinson, M.L.N. (eds) *Sustaining Earth: Response to the Environmental Threat*, London: Macmillan, Ch. 9, pp. 97–107.

Goudie, A. and Viles, H. (1997) *The Earth Transformed: an Introduction to Human Impacts on the Environment*, Oxford: Blackwell.

Gunn, C. (1994) *Tourism Planning: Basic, Concepts, Issues*, 2nd edn, Washington: Taylor and Francis.

Hall, C. and Lew, A. (eds) (1998) *Sustainable Tourism: a Geographical Perspective*, Harlow: Addison Wesley Longman.

Hall, D. and Kinnaird,V. (1994) 'Ecotourism in Eastern Europe', in Cater, E. and Lowman, G. (eds) *Ecotourism: a Sustainable Option*, Chichester: Wiley, Ch. 7, pp. 111–36.

Hardin, G. (1968) 'The Tragedy of the Commons', *Science*, 162: 1243–8.

Harrison, D. (1996) 'Sustainability and Tourism: Reflections from a Muddy Pool' in Briguglio, L., Archer, B., Jafari, J. and Wall, G. (eds) *Sustainable Tourism in Islands and Small States: Issues and Policies*, London: Pinter, Ch. 6, pp. 69–89.

Harrison, D. (1998) 'Whales Under Stress as Man Crowds the Sea', the *Observer*, 18 October, p. 7.

Hawkins, R. (1997) 'Green Labels for the Travel and Tourism Industry – A Beginner's Guide', *Insights*, July, pp. A11–A15, London: English Tourist Board.

Hertsgaard, M. (1999) *Earth Odyssey*, London: Abacus.

Holden, A. (1991) 'Asian Dynasty', *Leisure Management*, 11(12): 29–30.

Holden, A. (1998) 'The Use of Skier Understanding in Sustainable Management in the Cairngorms', *Tourism Management*, 19(2): 145–52.

Holden, A. and Kealy, H. (1996) 'A Profile of UK Outbound Environmentally Friendly Tour Operators?', *Tourism Management*, 17(1): 60–4.

Holdgate, M. (1990) 'Changes in Perception', in Angell, D.J.R., Comer, J.D. and Wilkinson, M.L.N. (eds) *Sustaining Earth*, Basingstoke: Macmillan, Ch. 8, pp. 76–96.

Holloway, C. (1998) *The Business of Tourism*, 5th edn, Harlow: Addison Wesley Longman.

Hooker, C.A. (1992) 'Responsibility, Ethics and Nature', in Cooper, D.E. and

Palmer, J.A. (eds) *The Environment in Question: Ethics and Global Issues*, London: Routledge, Ch. 11, pp. 147–64.

House, J. (1997) 'Redefining Sustainability: a Structural Approach to Sustainable Tourism', in Stabler, M. (ed.) *Tourism and Sustainability: Principles to Practice*, Wallingford: CAB International, Ch. 6, pp. 89–104.

Hudman, E. (1991) 'Tourism's Role and Response to Environmental Issues and Potential Future Effects', *Revue de Tourisme (The Tourist Review)*, 4: 17–21.

Hunter, C. (1996) 'Sustainable Tourism as an Adaptive Paradigm', *Annals of Tourism Research*, 24(4): 850–67.

Hunter, C. and Green, H. (1995) *Tourism and the Environment: A Sustainable Relationship?*, London: Routledge.

IFTO (1994) *Planning for Sustainable Tourism: The Ecomost Project*, Lewes: International Federation of Tour Operators.

Iso-Ahola, E.S. (1980) *The Social Psychology of Leisure and Recreation*, Iowa: Wm. C. Brown.

Ittleson, W.H., Franck, K.A. and O'Hanlon, T.J. (1976) 'The Nature of Environmental Experience' in Wagner, S., Cohen, B.S. and Kaplan, B. (eds) *Experiencing the Environment*, New York: Plenum Press, Ch. 9, pp. 187–206.

Jamrozy, U. and Uysal, M. (1994) in Uysal, M. (ed.) *Global Tourist Behaviour*, New York: The Haworth Press, Ch. 6, pp. 135–60.

Jenner, P. and Smith, C. (1992) *The Tourism Industry and the Environment*, London: The Economist Intelligence Unit.

Keefe, J. (1995) 'Water Fights', *Tourism in Focus*, 17: 8–9.

Kirkby, S.J. (1996) 'Recreation and the Quality of Spanish Coastal Waters', in Barke, M., Towner, J. and Newton, M.T. (eds) *Tourism in Spain: Critical Issues*, Wallingford: CAB International, Ch. 8, pp. 190–211.

Klemm, M. (1992) 'Sustainable Tourism Development: Languedoc and Roussillon', *Tourism Management*, June, 169–80.

Krippendorf, J. (1987) *The Holiday Makers*, Oxford: Heinemann.

Lanjouw, A. (1999) 'Mountain Gorilla Tourism in Central Africa', *owner-mtn-forum@igc.apc.org*.

Law, C. (1993) *Urban Tourism: Attracting Visitors to Large Cities*, London: Mansell.

Laws, E. (1991) *Tourism Marketing*, Cheltenham: Stanley Thornes.

Lea, J.P. (1993) 'Tourism Development Ethics in the Third World', *Annals of Tourism Research*, 20(4): 701–15.

Lechte, J. (1994) *Fifty Key Contemporary Thinkers: From Structuralism to Postmodernity*, London: Routledge.

Lěncek, L. and Bosker, G. (1998) *The Beach: the History of Paradise on Earth*, London: Secker and Warburg.

Leopold, A. (1949) *A Sand County Almanac*, Oxford: Oxford University Press.

Lickorish, L.J. and Jenkins, C.L. (1997) *An Introduction to Tourism*, Oxford: Butterworth-Heinemann.

Lovelock, J. (1979) *Gaia: a New Look at Life on Earth*, Oxford: Oxford University Press.

MacCannell, D. (1976) *The Tourist: a New Theory of the Leisure Class*, New York: Schocken Books.

MacCannell, D. (1989) *The Tourist*, 2nd edn, London: Macmillan.

MacCannell, D. (1992) *Empty Meeting Grounds: the Tourist Papers*, London: Routledge.

McCool, S.F. (1996) 'Limits of Acceptable Change: a Framework for Managing National Protected Area: Experiences from the United States', paper presented at the *Workshop in Impact Management in Marine Parks*, Kuala Lumpur, Malaysia, August 13–14.

Mackay, A. (1994) 'Eco Tourists Take Over', *The Times*, London, 17 February.

McLaren, D. (1998) *Rethinking Tourism and Ecotravel*, Connecticut: Kumarian Press.

Malone, P. (1998) 'Pollution Battle Takes to the Skies', the *Observer*, London, 8 November.

Marcel-Thekaekara, M. (1999) 'Poor relations', the *Guardian*, Saturday Review Section, 27 February, p. 3.

Martin, A. (1997) *Tourism, the Environment and Consumers*, paper given at 'The Environment Matters' conference, Glasgow, April 30.

Maslow, A.H. (1954) *Motivation and Personality*, New York: Harper.

Mason, P. and Mowforth, M. (1996) 'Codes of Conduct in Tourism', *Progress in Tourism and Hospitality Research*, 2(2): 151–68.

Mathieson, A. and Wall, G. (1982) *Tourism: Economic, Physical and Social Impacts*, Harlow: Longman.

Middleton, V. (1988) *Marketing in Travel and Tourism*, Oxford: Heinemann.

Mieczkowski, Z. (1995) *Environmental Issues of Tourism and Recreation*, Lanham, MD: University Press of America.

Mill, R.C. and Morrison, A.M. (1992) *The Tourism System: an Introductory Text*, 2nd edn, New Jersey: Prentice Hall.

Milne, S. (1988) 'Pacific Tourism: Environmental impacts and their management', paper presented to the Pacific Environmental Conference, London, 3–5 October.

Mishan, E.J. (1969) *The Costs of Economic Growth*, Harmondsworth: Penguin.

Monbiot, G. (1995) 'No man's land', *Tourism in Focus*, 15: 10–11, London: Tourism Concern.

Moynahan, B. (1985) *The Tourist Trap*, London: Pan Books.

Mowforth, M. and Munt, I. (1998) *Tourism and Sustainability: New Tourism in the Third World*, London: Routledge.

Murphy, P. (1985) *Tourism: a Community Approach*, London: Routledge.

Murphy. P. (1994) 'Tourism and Sustainable Development', in Theobald, W. (1994) *Global Tourism: the Next Decade*, Oxford: Butterworth-Heinemann, Ch. 19, pp. 274–90.

Nash, D. (1979) 'The Rise and Fall of an Aristocratic Tourist Culture', *Annals of Tourism Research*, Jan/March, pp. 63–75.

Nash, R.F. (1989) *The Rights of Nature: a History of Environmental Ethics*, Wisconsin: The University of Wisconsin Press.

New Internationalist (1999) 'The Radical Twentieth Century', No. 309, Oxford.

Nicholson-Lord, D. (1993) 'Mass tourism is blamed for paradise lost in Goa', the *Independent*, January 27, pp. 10–11.

O'Reilly (1986) 'Tourism Carrying Capacity: Concepts and Issues', *Tourism Management*, 8(2): 254–8.

O'Riordan, T. (1981) *Environmentalism*, 2nd edn, London: Pion.

Osborn, D. and Bigg, T. (1998) *'Earth Summit II: Outcomes and Analysis'*, London: Earthscan.

Page, J. (1999) *Travel and Tourism*, the *Guardian* (weekend section), November 6, p. 102.

Page, S. (1995) *Urban Tourism*, London: Routledge.

Panos (1995) 'Ecotourism: Paradise Gained, or Paradise Lost', *Panos Media Briefing*, 14: 1–15.

Parviainen, J., Pöysti, E. and Kehitys, S. (1995) *Towards Sustainable Tourism in Finland*, Helsinki: Finnish Tourist Board.

Pattullo, P. (1996) *Last Resorts: the Cost of Tourism in the Caribbean*, London: Cassell.

Pearce, D., Markandya, A. and Barbier, E. B. (1989) *Blueprint for a Green Economy*, London: Earthscan Publications.

Pearce, D. (1993a) *Economic Values and the Natural World*, London: Earthscan Publications.

Pearce, P. (1993b) 'Fundamentals of Tourist Motivation', in Pearce, D.G. and Butler, R.W. (eds), *Tourism Research: Critiques and Challenges*, London: Routledge, Ch. 7, pp. 113–34.

Pearce, P. (1988) *The Ulysses Factor: Evaluating Visitors in Tourist Settings*, New York: Springer-Verlag.

Pepper, D. (1993) *Eco-socialism: From Deep Ecology to Social Justice*, London: Routledge.

Pepper, D. (1996) *Modern Environmentalism: an Introduction*, London: Routledge.

Pi-Sunyer, O. (1996) 'Tourism in Catalonia' in Barke, M., Towner, J. and Newton, M.T. (eds) *Tourism in Spain: Critical Issues*, Wallingford: CAB International, Ch. 10, pp. 231–64.

Plog, S. (1974) 'Why Destination Areas Rise and Fall', *Cornell Hotel and Restaurant Quarterly*, Nov., pp. 13–16.

Ponting, C. (1991) *A Green History of the World*, London: Sinclair-Stevenson.

Poon, A. (1993) *Tourism, Technology and Competitive Strategies*, Wallingford: CAB International.

Porritt, J. (1984) *Seeing Green: the Politics of Ecology Explained*, Oxford: Basil Blackwell.

Reid, D. (1995) *Sustainable Development: an Introductory Guide*, London: Earthscan.

Reuters (1999) 'Cape to Seek Sex Tourists', the *Guardian*, London, September 20, p. 12.

Richardson, D. (1997) 'The Politics of sustainable development', in Baker, S., Kousis, M., Richardson, D. and Young, S. (1997) *The Politics of Sustainable Development: Theory, Policy and Practice within the European Union*, London: Routledge, Ch. 1, pp. 43–60.

Rogers, P. and Aitchison, A. (1998) *Towards Sustainable Tourism in the Everest Region of Nepal*, Kathmandu: International Union for Conservation (IUCN).

Roe, D., Leader-Williams N. and Dalal-Clayton, B. (1997) *Take Only Photographs: Leave Only Footprints*, London: International Institute for Environment and Development.

Roussopoulos, D.I. (1993) *Political Ecology*, Montreal: Black Rose Books.

Ryan, C. (1991) *Recreational Tourism: a Social Science Perspective*, London: Routledge.

Salem, N. (1995) 'Water Rights', *Tourism in Focus*, 17: 4–5.

Scottish Enterprise (1995) *A Study of Activity Holidays*, Edinburgh: Scottish Tourist Board.

Scottish National Heritage (1993) *Sustainable Development and the Natural Heritage: the SNH approach*, Perth.

Scottish Office (1996) *National Planning Policy Guidelines for Skiing*, Edinburgh: Scottish Office.

Scruton, R. (1999) 'Please Don't Give Me Legal Rights', *Evening Standard*, London, February 12, p. 31.

Shackley, M. (1995) 'The Future of Gorilla Tourism in Rwanda', *Journal of Sustainable Tourism*, 3(2): 61–72.

Shackley, M. (1996) *Wildlife Tourism*, London: International Thomson Business Press.

Sharpley, R. (1994) *Tourism, Tourists and Society*, Huntingdon: Elm Publications.

Shaw, S. (1993) *Transport: Strategy and Policy*, Oxford: Blackwell.

Short, J.R. (1991) *Imagined Country: Society, Culture and Environment*, London: Routledge.

Simons, P. (1988) 'Après ski le deluge', *New Scientist*, 1: 46–9.

Simmons, I.G. (1993) *Interpreting Nature: Cultural Constructions of the Environment*, London: Routledge.

Simmons, M. and Harris, R. (1995) 'The Great Barrier Reef Marine Park', in Harris, R. and Leiper, N. (eds) *Sustainable Tourism: an Australian Perspective*, Oxford: Butterworth-Heinemann.

Sinclair, T.M. (1991) 'The Tourism Industry and Foreign Exchange Leakages in a Developing Country: the Distribution of Earnings from Safari and Beach Tourism in Kenya', in Sinclair, T.M. and Stabler, M.J. (eds) *The Tourism*

Industry: an International Analysis, Wallingford: CAB International, Ch. 10, pp. 185–204.

Sinclair, T.M. and Stabler, M. (1997) *The Economics of Tourism*, London: Routledge.

Singer, P. (1993) *Practical Ethics*, 2nd edn, Cambridge: Cambridge University Press.

Slattery, M. (1991) *Key Ideas in Sociology*, Walton-on-Thames, Surrey: Thomas Nelson and Sons.

Smith, C. and Jenner, P. (1992) 'The Leakage of Foreign Exchange Earnings from Tourism', in Economist Intelligence Unit, *Travel and Tourism Analyst* (3), London: Economist Intelligence Unit.

Smout, C. (1990) *The Highlands and the Roots of the Green Consciousness*, Perth: Scottish National Heritage.

Somerville, C., Rickmers, R.W. and Richardson, E.C. (1907) *Ski Running*, 2nd edn, London: Horace Cox.

Sparrowhawk, J. and Holden, A. (1999) 'Human Development: the Role of Tourism Based NGOs in Nepal', *Tourism Recreation Research*, 24(2): 37–44.

Stabler, M. and Goodall, B. (1996) 'Environmental Auditing in Planning for Sustainable Island Tourism', in Briguglio, L., Archer, B., Jafari, J. and Wall, G. (eds) *Sustainable Tourism in Islands and Small States: Issues and Policies*, London: Pinter, Ch. 12, pp. 170–96.

Stabler, M. and Sinclair, M.T. (1997) *The Economics of Tourism*, London: Routledge.

Stabler, M. (forthcoming) 'Local community influences on the management and conservation of natural environments for leisure and tourism', Discussion Paper in Urban and Regional Economics, Department of Economics, University of Reading.

Stevens, T. (1987) 'Wigan Pier', *Leisure Management*, 7(6): 31–4.

Stone, C.D. (1993) *The Gnat Is Older than Man: Global Environment and Human Agenda*, Princeton: Princeton University Press.

Swarbrooke, J. and Horner, S. (1999) *Consumer Behaviour in Tourism*, Oxford: Butterworth-Heinemann.

Todd, S.E. and Williams, P.W. (1996) 'Environmental Management System Framework for Ski Areas', *Journal of Sustainable Tourism*, 4(3): 147–73.

Tourism Industry Association of Canada (1995) *Code of Ethics and Guidelines for Sustainable Tourism*, Ottawa: Tourism Industry Association of Canada.

Tourism Planning and Research Associates (1995) *The European Tourist*, 8th edn, London: Tourism Planning and Research Associates.

Towner, J. (1996) *An Historical Geography of Recreation and Tourism in the Western World: 1540–1940*, Chichester: Wiley.

TTG (1999) 'No Waste of Space', *Travel Trade Gazette*, p. 19, Special Edition (Day 3), World Travel Market, London.

Tunstall, S.M. and Penning-Rowsell, E. C. (1998) 'The English Beach: Experiences and Values', *The Geographical Journal*, 163(3): 319–32.

Turner, L. and Ash, J. (1975) *The Golden Hordes: International Tourism and the Pleasure Periphery*, London: Constable.

Turner, K.R., Pearce, D. and Bateman, I. (1994) *Environmental Economics: an Elementary Introduction*, Hemel Hempstead: Harvester Wheatsheaf.

United Nations Environment Programme (1995) *Environmental Codes of Conduct for Tourism: Technical Report*, No. 29, UNEP, Paris.

Urry, J. (1990) *The Tourist Gaze: Leisure and Travel in Contemporary Societies*, London: Sage.

Urry, J. (1995) *Consuming Places*, London: Routledge.

Vardy, P. and Grosch, P. (1999) *The Puzzle of Ethics*, London: Fount.

Veblen, T. (1899) *The Theory of the Leisure Class*, republished in 1994 (3rd edition) by Penguin, London.

Vidal, J. (1994) 'Money for Old Hope', the *Guardian*, London, January 7, pp. 14–15.

Visser, N. and Njuguna, S. (1992) 'Environmental Impacts of Tourism on the Kenya coast', *Industry and Environment*, 15(3): 42–51,United Nations Environment Programme, Paris.

Waldeback, K. (1995) *Beneficial Environmental Sustainable Tourism*, Vanuatu: BEST.

Ward, L. (1998) 'Quality of life gets a higher profile', the *Guardian*, November 24, p. 2.

Wathern, P. (ed.) (1988) *Environmental Impact Assessment: Theory and Practice*, London: Routledge.

Watson-Smyth, K. (1999) 'Britain in 2010: Rich but Far too Stressed to Enjoy It', the *Independent*, London, September 15, p. 8.

WCED (1987) *Our Common Future*, Oxford: Oxford University Press.

Wearing, S. and Neil, J. (1999) *Ecotourism: Impacts, Potentials and Possibilities*, Oxford: Butterworth-Heinemann.

Weston, J. (ed.) (1997) *Planning and Environmental Impact Assessment*, Harlow: Addison Wesley Longman.

Wheeller, B. (1993a) 'Sustaining the ego?', *Journal of Sustainable Tourism*, 1(2): 23–9.

Wheeller, B. (1993b) 'Willing victims of the ego trap', *Tourism in Focus*, 9: 10–11.

Whitelegg, J. (1999) *Air Transport and Global Warming*, www.gn.apc.org/sgr/kyoto/jw.html.

Wight, P. (1994) 'Environmentally Responsible Marketing of Tourism', in Cater, E. and Lowman, G. (eds) *Ecotourism: A Sustainable Option*, Chichester: Wiley, Ch. 3, pp. 39–53.

Wight, P. (1998) 'Tools for Sustainability Analysis in Planning and Managing Tourism and Recreation in a Destination', in Hall, C. and Lew, A. (eds) *Sustainable Tourism: a Geographical Perspective*, Harlow: Addison Wesley Longman, Ch. 7, pp. 75–91.

Williams, P.W. and Gill, A. (1994) 'Tourism Carrying Capacity Management Issues', in Theobald, W. (ed.) *Global Tourism: the Next Decade*, Oxford: Butterworth-Heinemann, Ch. 12, pp. 174–87.

Williams, S. (1998) *Tourism Geography*, London: Routledge.

Willis, I. (1997) *Economics and the Environment: a Signalling and Incentives Approach*, St. Leonards: Allen and Unwin.

Wood, K. and House, S. (1991) *The Good Tourist*, London: Mandarin Paperbacks.

Wood, L. (1998) 'Quality of life gets a higher profile', the *Guardian*, November 24, p. 2.

World Guide (1997/8) *The World Guide: A View from the South*, Oxford: New Internationalist Publications.

World Tourism Organisation (1991) *International Conference on Travel and Tourism Statistics*, Madrid: WTO.

World Tourism Organisation (1992) *Tourism Carrying Capacity: Report on the Senior-Level Expert Group Meeting held in Paris, June 1990*, Madrid: WTO.

World Tourism Organisation (1998a) *Tourism: 2020 Vision (Executive Summary Updated)*, Madrid: WTO.

World Tourism Organisation (1998b) 'Global Warming', *WTO News*, p. 6.

World Tourism Organisation (1999a) *Tourism Highlights: 1999*, Madrid: WTO.

World Tourism Organisation (1999b) 'Global Code of Ethics for Tourism', www.world-tourism.org/pressrel/CODEOFE.htm.

World Travel and Tourism Council (1999) *WTTC: Key Statistics*, www.wttc.org/economic_research/keystats.htm.

Worthington, S. (1999) 'Green Slogans Are Whitewash', *Evening Standard*, London, 7 October, p. 23.

Zimmermann, F.M. (1995) 'The Alpine Region: Regional Restructuring Opportunities and Constraints in a Fragile Environment', in Montanari, A. and Williams, A.M. (eds) *European Tourism: Regions, Spaces and Restructuring*, Chichester: Wiley.

Index